International Scientific Conference of Communications, Information, Electronic and Energy Systems (CIEES 2021)

IOP Conference Series: Materials Science and Engineering
Volume 1216

Ruse, Bulgaria
25 - 27 November 2021

ISBN: 978-1-7138-4675-8
ISSN: 1757-8981

Printed from e-media with permission by:

Curran Associates, Inc.
57 Morehouse Lane
Red Hook, NY 12571

Some format issues inherent in the e-media version may also appear in this print version.

Institute of Physics
Dirac House, Temple Back
Bristol BS1 6BE UK

Phone: 44 1 17 929 7481
Fax: 44 1 17 920 0979

techtracking@iop.org

Additional copies of this publication are available from:

Curran Associates, Inc.
57 Morehouse Lane
Red Hook, NY 12571 USA
Phone: 845-758-0400
Fax: 845-758-2633
Email: curran@proceedings.com
Web: www.proceedings.com

TABLE OF CONTENTS

Preface to the International Scientific Conference on Communications, Information, Electronic and Energy Systems (CIEES'2021)

The International Scientific Conference on Communications, Information, Electronic and Energy Systems (CIEES'2021) was held virtually in Ruse, Bulgaria from 25 to 27 November, 2021. The CIEES'2021 was organized by the Association CIEES with the following partners: University of Ruse (Bulgaria), Politehnica University of Bucharest (Romania), the Union of Scientists in Ruse (Bulgaria) and Encon Services Ltd (Bulgaria).

In light of the current situation related to the coronavirus (COVID-19) and regarding to the internal government regulations from the Bulgarian Ministry of Health, the Organizing Committee took the decision to turn CIEES'2021 in a virtual format – online conference on the same period.

The goal of the CIEES'2021 conference is to provide an international forum for researchers, academics, people in industry, and students to address recent research results and to present and discuss their ideas, theories, technologies, systems, tools, applications, work in progress and experiences on all theoretical and practical issues arising in telecommunications, electronics, signal processing and applications, information technologies and energy.

The technical program of this year's edition of the conference consists of three distinguished keynotes by world-renowned experts with their lectures as follows:

- Prof. Stefan Ditchev DSc, University of Food Technologies – Plovdiv, Bulgaria, the title of the lecture "*The heat pump systems – an option for energy efficiency and innovation of technologies*";
- Assist. Prof. Dr. Frank Grabner, IMG Electronic & Power Systems GmbH, Nordhausen, Germany, the title of the lecture "*New EMC effects with multi-layered type of EM shield*";
- Prof. Dr. Felipe L. Peñaranda-Foix, Universidad Politécnica de Valencia (UPV), Valencia, Spain, the title of the lecture "*Material electromagnetic characterization*".

During the online conference, were used the Zoom Meeting platform for the opening ceremony, all conference sessions, and meetings. In advance, the Conference Organizing Committee of CIEES 2021 organized test sessions for participants, Conference chairs, and plenary lecturers. All authors prepared a video presentation of up to 12 minutes that was played on stream with authors participating in the discussion during the session. All of the video presentations are available on the conference YouTube© channel.

The conference programme consists 3 special sessions and 7 technical sessions. The time slot for presentation and Q&A was 15 minutes. Questions were asked in the chat section of the zoom channel or by raised hand when the video presentation is finished. All Q&A were made face-to-face in the Zoom channel, between the CIEES Host (technical moderator), Session Chair, and participants.

Among the submitted manuscripts, 20 full papers passed the review process for publication in IOP Conference Series: Materials Science and Engineering. The review process was conducted according to the review policy of the IOP conference series, involving national and international reviewers - each paper was reviewed by two reviewers. During the discussion phase, International Scientific Committee members and the Programme Committee members discussed the reviews in detail. The rebuttals from the authors were also seriously considered. For consistency across tracks, final decisions on papers were made at the face-to-face Programme Committee meeting between the Organizing Committee chairs and the Programme Committee chairs.

Finally, we are grateful to all members of the International Scientific Committee, for providing insightful guidelines and support in preparing and organizing the conference, and for the highly professional work done on all aspects of the local organization of the Conference.

We thank all authors and participants for their contributions.

CIEES is an annual conference. More information about the conference and the forthcoming events, please visit the website: http://ciees.eu.

Teodor ILIEV
General Conference Chair
tiliev@uni-ruse.bg

Ivaylo STOYANOV
General Conference Chair
stoyanov@uni-ruse.bg

CIEES 2021 Organizing Committee

General Conference Chairs:
Teodor ILIEV (Bulgaria)
Ivaylo STOYANOV (Bulgaria)

Honorary Chair:
Hristo BELOEV (Bulgaria)

Technical Programme Chair:
Grigor MIHAYLOV (Bulgaria)

Scientific Committee Chair:
Bogdan POPA (Romania)

Awards Committee Chair:
Iliya ILIEV (Bulgaria)

Guest Editors:
Iliya ILIEV (Bulgaria)
Renato FILJAR (Croatia)

Web Chair:
Yordan DOYCHINOV (Bulgaria)

Programme Committee
Chair: Filip TSVETANOV (Bulgaria)
Gani BALBAYEV (Kazakhstan)
Goran DJORDJEVIC (Serbia)
Dejan MILIC (Serbia)
Elena IVANOVA (Bulgaria)
Jacob FANTIDIS (Greece)
Panagiotis KOGIAS (Greece)

CIEES 2021 International Scientific Committee

Chair: Renato FILJAR (Croatia)
Felipe PENARANDA-FOIX (Spain)
Selahattin KOSUNALP (Turkey)
Filip TSVETANOV (Bulgaria)
Ivanka GEORGIEVA (Bulgaria)
Panagiotis KOGIAS (Greece)
Svilen RACHEV (Bulgaria)
Bogdan POPA (Romania)
Iliya ILIEV (Bulgaria)
Frank GRÄBNER (Germany)
Dragos-Alexandru ANDRIOAIA (Romania)
Ozgur CULFA (USA)
Lawal Mohammed BELLO (Nigeria)
Gani BALBAYEV (Kazakhstan)
Goran DJORDJEVIC (Serbia)
Hristo VALCHANOV (Bulgaria)
Angel TERZIEV (Bulgaria)
Mehmet Emin BAKIR (UK)
Nenad SIKIRICA (Croatia)
Alexander GENBACH (Kazakhstan)
Petru-Gabriel PUIU (Romania)
Veneta ALEKSIEVA (Bulgaria)
Darko HULJENIC (Croatia)
Eliza TICA (Romania)
Jacob FANTIDIS (Greece)
Dejan MILIC (Serbia)
Oliver JUKIĆ (Croatia)
Jonatan LERGA (Croatia)
Kristijan LENAC (Croatia)
Darko SPOLJAR (Croatia)
Marko HORVAT (Croatia)
Peter BRIDA (Slovakia)
Sandi LJUBIC (Croatia)
Boris EVSTATIEV (Bulgaria)
Lorant SZOLGA (Romania)
Ogunjo Samuel TOLUWALOPE (Nigeria)
Assel MUKASHEVA (Kazakhstan)
Teodora HRISTOVA (Bulgaria)
Krasimira IVANOVA (Bulgaria)

Peer review declaration

All papers published in this volume of *IOP Conference Series: Materials Science and Engineering* have been peer reviewed through processes administered by the Editors. Reviews were conducted by expert referees to the professional and scientific standards expected of a proceedings journal published by IOP Publishing.

- **Type of peer review: Single-blind**
- **Conference submission management system: Microsoft CMT - https://cmt3.research.microsoft.com/**
- **Number of submissions received: 28**
- **Number of submissions sent for review: 28**
- **Number of submissions accepted: 17**
- **Acceptance Rate (Number of Submissions Accepted / Number of Submissions Received X 100): 60.71%**
- **Average number of reviews per paper: 2**
- **Total number of reviewers involved: 31**
- **Any additional info on review process:** All manuscripts submitted to the CIEES'2021 conference were single-blind peer-reviewed and are carried out to meet the scientific criteria of novelty and academic excellence. The review process was conducted according to the review policy of IOP conference series - each paper was reviewed by at least two reviewers involving both national and international reviewers. During the discussion phase, International Scientific Committee members and the Programme Committee members discussed the reviews in detail. The rebuttals from the authors were also seriously considered. The final decision to accept a paper was entirely based on quality and not its length.
- **Contact person for queries:** Assoc. Prof. Teodor Iliev, PhD, Department of Telecommunications, University of Ruse, Bulgaria, E-mail: tiliev@uni-ruse.bg

Cloud computing security requirements: A Review

G R Tsochev[1], R I Trifonov[1]

[1] Department of Information Technologies in Industry, Technical University of Sofia, Bulgaria

E-mail: gtsochev@tu-sofia.bg, r_trifonov@tu-sofia.bg

Abstract. Cloud computing is a new technology that is undergoing tremendous development today. People who use it are not able to separate the reasonable from the unreasonable arguments that come with the security requirements in the cloud. The claim that cloud computing is hereditarily insecure is as absurd as the claim that cloud computing does not create new security problems. Cloud computing is a way to dynamically increase resources without the need for in-depth knowledge of a brand new infrastructure, without training new workers or designing new software solutions. The article aims to analyse the different cloud security issues and models of cloud architectures. Some of the main problems with security in virtualization, concerns about storing data in the cloud and the assessment of risk tolerance in cloud computing are presented. Legal and regulatory issues for the protection of personal data are addressed.

1. Introduction

Cloud Computing is the most important trend in the IT Industry (figure 1). Even the biggest critics seem to agree that – in spite of some over-zealous marketeers – Cloud Computing is one of the most important paradigm shifts of the past decades. But what is it all about? Where did it come from? And what's to be expected? There are probably as many definitions of Cloud Computing as there are self-acclaimed Cloud Specialist.

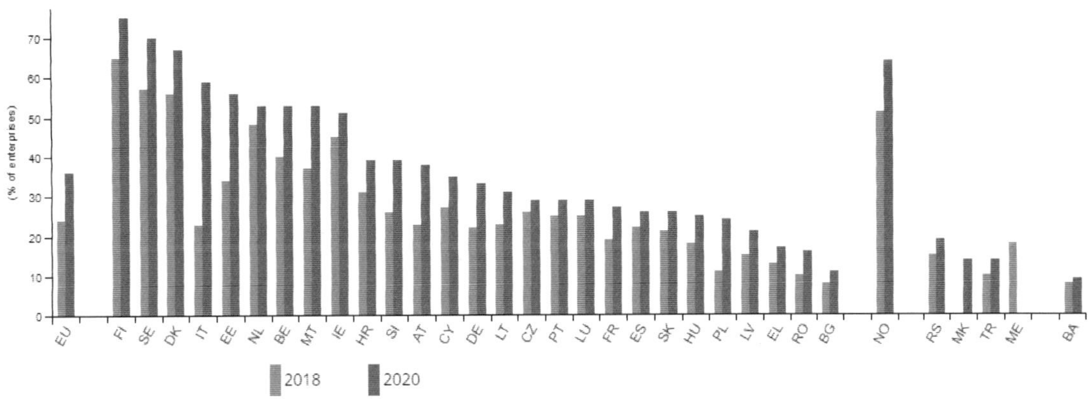

Figure 1. Use of cloud computing services in EU [1]

IOP Conf. Series: Materials Science and Engineering 1216 (2022) 012001 doi:10.1088/1757-899X/1216/1/012001

The truth is that the cloud computing concept is really simply, it is has been simple for decades. Have you ever been heard the sayings "Two heads thinks better than one.", "Nobody knows everything but everybody knows something." and so on and so on? The technical definition could sound like "Cloud computing is Internet-based computing, whereby shared resources, software, and information are provided to computers and other devices on demand, like the electricity grid." [2] or "Cloud computing is the utilization of many servers housed in many different locations and interconnected by high speed networks." [3] although the marketing and sales guys are defining it as a new way of delivering IT services that are cheaper, flexible and really easy to use. As you may already think the truth is somewhere between. The main specifications of the cloud computing are lower cost, flexibility, scalability and "pay-per-view" or usage on demand. Cloud computing is on-demand access to virtualized IT resources that are housed outside of your own data center, shared by others, simple to use, paid for via subscription, and accessed over the Web. While this does reflect a common experience, in reality it is a fairly limiting definition. For example, what about the requirement that everything is provided ''as a service over the Internet?'' That might seem attractive at first, yet it does not allow for the reality that there are many applications—for sound, unemotional, pragmatic considerations—that will require a private deployment (off the Internet). A team at the National Institute of Standards and Technology (NIST) has been doing some very good work to bring some order to these discussions [4]. Here is their working definition:

"Cloud computing is a model for enabling convenient, on-demand network access to a shared pool of configurable computing resources (e.g., networks, servers, storage, applications, and services) that can be rapidly provisioned and released with minimal management effort or service provider interaction. This cloud model promotes availability and is composed of five essential characteristics, three delivery models, and four deployment models." [5]

The most important is to understand that cloud computing is not a technology, it is a concept. It is a concept that could be used in many different ways, cloud computing is Gmail, amazon, Google, SaleForge, torrents, etc. This concept simply means that if you connect two computers and make them work as one they will have more computing power then every one of them separated. Now imagine if you connect thousands (figure 2).

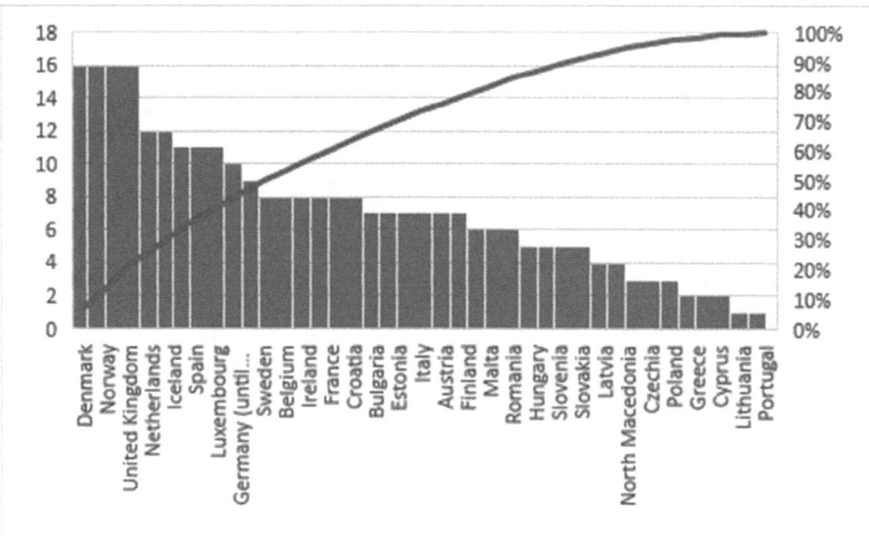

Figure 2. Problems experienced when using cloud services [1]

Cloud services have gradually become a key feature of modern computers, smartphones and tablets [6]. Each leading company in the industry offers consumers its own offer in this segment (figure 3), which significantly intensifies the competition in it. Cloud technologies are a flexible, highly efficient and proven platform for providing IT services over the Internet. However, their increasing use and the provision of cloud services by providers creates a number of risks associated with ensuring the protection of information.

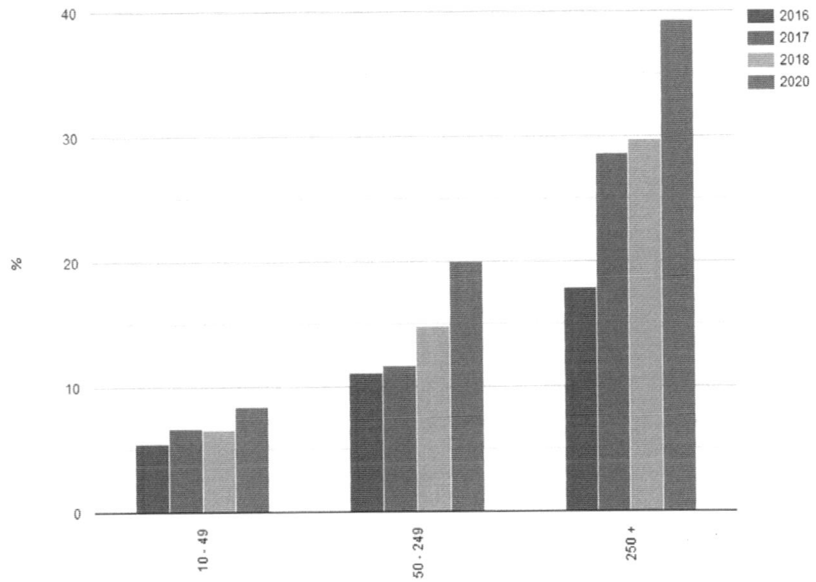

Figure 3. Enterprises using paid computer services in the cloud [7]

2. Security architecture

In order to avoid ineffective and expensive operation, broken governance, hard to automate controls or procedures and poor security, sound foundations of principles, structure and methodology must be established.

Systems engineering is a methodology for achieving integration by viewing the collective of components as a hole rather than an assembly. Each component should be designed in a way that will interoperate with the rest. Figure 4 shows the scope systems engineering.

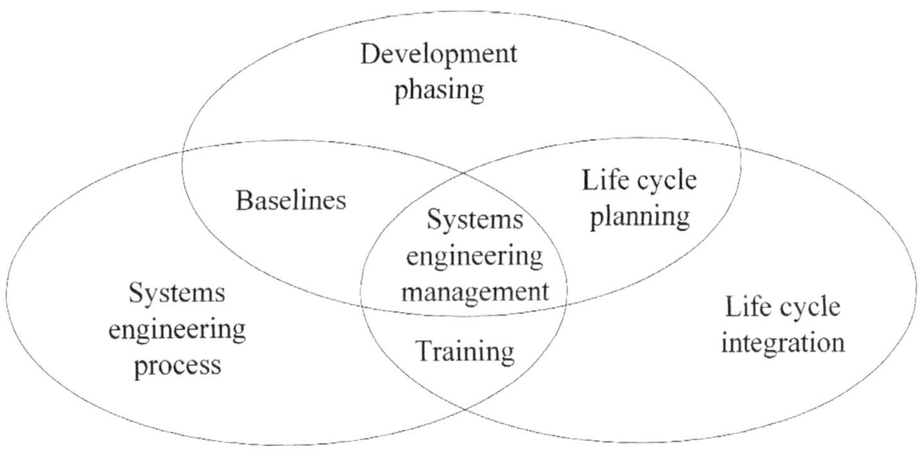

Figure 4. Systems engineering management activities [8]

In IT architecture, the focus is on designing the infrastructure of individual components. The primary objective is to achieve an efficient structure that meets the needs or mission of an organization over a long tern in a sustainable manner. It should present stability, room for continual innovation, support business goals and help reduce costs [8].

Security Architecture has been influenced by architectural approaches, engineering and models. The System Security Engineering Capability Maturity Model (SSE-CMM) created in early 2000s emphasizes on the importance of practicing security engineering and many new models are based on it. These type of models can serve as reference models for security engineering, security architecture, security operations as well as cloud computing [8]. Some noteworthy to mention models are:

ISO 27001 - ISO 27006: The International Standards for security. They cover management, best practices, requirements and techniques.

European Network and Information Security Agency (ENISA): The European cyber security agency's recommendations for security issues when adopting cloud computing.

Information Technology Infrastructure Library (ITIL): catered toward business needs it focuses on IT service management and is structured around service life cycles and practices.

Control Objectives for Information and related Technology (COBIT): a set of generally accepted best practices for governance and control [9].

The National Institute for Standards and Technology (NIST) covers standards and guidelines for non-government security engineering and architecture.

Security architecture can help bring together a common set of security requirements from multiple stakeholders and address them as a collective with a better solution than can be presented to them individually.

3. Some security problems

Restricting the freedom and creativity of users and creating a strong dependence on the cloud service provider is among the issues discussed in cloud computing. According to Richard Stallman, founder of the Free Software Foundation, "cloud computing threatens personal freedom because users provide their personal data to a third party." His suspicions are related to the imposition of certain systems and platforms by software companies on users. This is why Stallman expressed his view that cloud services will be difficult to enter in areas such as defence, government institutions, e-services, etc., where access to internal corporate networks must be strictly restrictive. However, practice shows that this is contrary to his understanding. Such examples are the US government, which switched to cloud services, the Ministry of Finance of the Republic of Bulgaria, which created a private cloud in 2012.

Another problem arising from the use of IT services in the cloud is related to data security. It means becoming highly dependent on the provider and the options it offers to protect data during transfer and storage. Many questions arise that are relevant to future hosting conditions: the possibility of information leaks to competing companies or malicious individuals, what mechanism to implement a possible change of provider, what will follow in the event of a system failure and others.

A service portfolio describes the services of the Cloud provider in terms of business value. It is a dynamic method used to govern investments in service management across the enterprise, in terms of financial values. With Service Portfolio Management (SPM), managers are able to assess the quality requirements and accompanying costs.

Figure 5 shows threats and their classification in cloud computing. The threats are generally grouped into five groups - network, data, related to virtualization, confidentiality and privacy and other common issues.

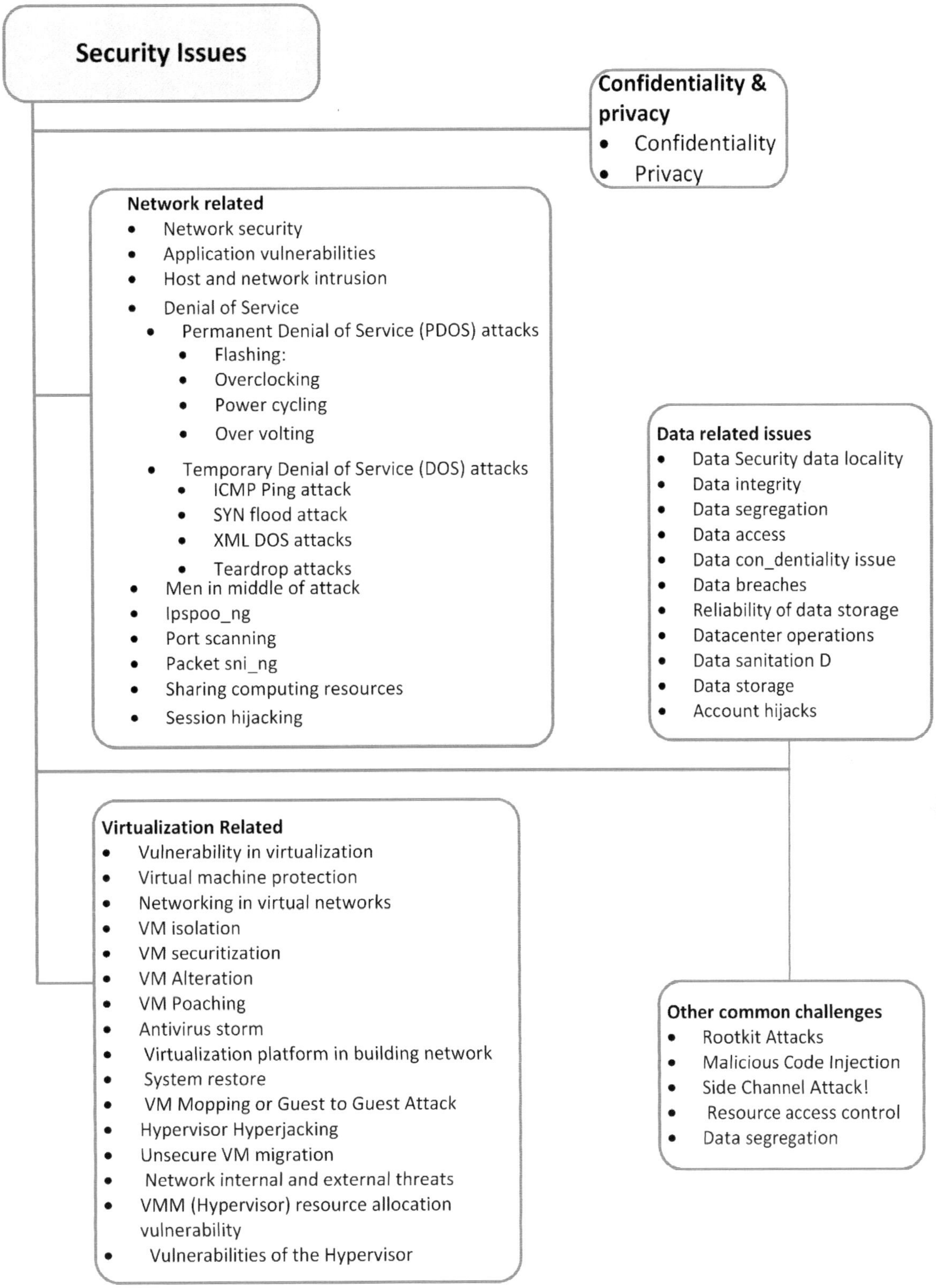

Figure 5. Appropriate security

4. Legal and regulatory issues for the protection

New laws are proposed all the time which can change the responsibilities of providers and customers and effect different parties under various scenarios. Legal issues can inevitably arise when dealing with data collection, storage or processing and they must be considered in order to be in legal compliance. Failure to do so can lead to potential fines by one or more government or industry regulatory bodies. The Data Protection Agency is the one of the government bodies who can make enterprises accountable for their actions in the European Union.

5. Risk Management

It is not always efficient to apply the same architecture and security controls for a low and a high risk environment (for example a banking application is not very likely to co-exist with a social-networking application in a public cloud). Risk can be defined as a function of threats that try to take advantage of vulnerabilities, divided by the countermeasures applied to protect our assets [8].

Risk= (Threats X Vulnerabilities / Countermeasures) X (Asset Value)

Thus a balance between the exposure side and cost side must be made. In order to implement a cost-effective security the risk must be determined according to an appropriate order of magnitude (figure 6).

However it is important to keep in mind that new security threats and vulnerabilities appear daily and more security layers means a better insurance.

Security must be part of the IT plans in the earliest stages in order to become business enabler.

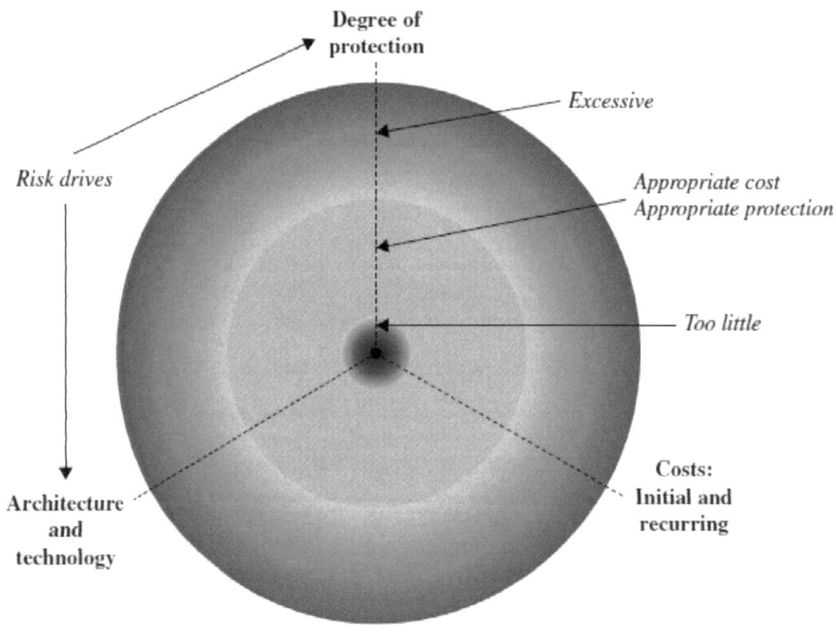

Figure 6. Appropriate security [8]

6. Conclusion and future work

Businesses and governments will continue to migrate to the cloud environment with the idea of reducing costs, improving efficiency and scalability, and reducing administrative difficulties. The delivery of cloud IT services saves time, money and improves productivity, but despite all the improvements that are in place, there are many risks to the security of user information that must be considered and addressed in a timely manner.

Security in the cloud is a vast topic that has been and will continue to be discussed many times. What can be concluded is that it provides a field for the development of many new technologies, new methods of protection. This is the theme in which most innovation is born and contributes to the advancement of technology. The advantages of the cloud are indisputable, and therefore more and more large companies are turning to it, which makes it an attractive prey for more and more hackers.

The analysis of cloud security will continue as part of the project "Possibility Investigation of Increasing the Cybersecurity of Systems in Industry 4.0 using Artificial Intelligence". It will be considered what basic methods of artificial intelligence can be applied to increase security or at least to detect anomalies in the context of cloud computing.

7. References

[1] Individuals - internet activities https://appsso.eurostat.ec.europa.eu/nui/show.do?Dataset = isoc_ci_ac_i&lang=en

[2] Jain A, Hada Ch and Buksh B 2016 Overview of Cloud Computing With Security Issues, Challenges & Countermeasures, *International Journal of Science, Technology and Management* (IJSTM vol 5 issue 5) ISSN: 2394-1537

[3] Talib A M, Atan R, Abdullah R and Azrifah M 2011 Cloud Zone: Towards an integrity layer of cloud data storage based on multi agent system architecture *2011 IEEE Conference on Open Systems* pp 127-132 doi: 10.1109/ICOS.2011.6079311

[4] Gavrilov G and Trajkovik V 2012 *Security and privacy issues and requirements for healthcare cloud computing*

[5] Jansen W A Cloud Hooks 2011 Security and Privacy Issues in Cloud Computing *In Proceedings of the 44th Hawaii International Conference on System Sciences 2011* (2)

[6] Chehlarova N and Miltchev R 2020 Development of Digital Competencies and Skills *The Field of Use of Cloud Services and Electronic Communication 5* pp 41-50

[7] Enterprises Using Paid Cloud Computing Services https://infostat.nsi.bg/infostat/pages/ reports/ result.jsf?x_2=778

[8] Vic Winkler J R 2011 *Securing the Cloud: Cloud Computer Security Techniques and Tactics* (Syngress Publishing)

[9] Mechtr M 2014 Virtual networked infrastructure provisioning in distributed cloud environments. Networking and Internet Architecture *Institut National des Télécommunications* NNT 2014 TELE0028

Acknowledgments

This research is realized and funded under the scientific-research project № КП-06-ПН47/27 “Possibility Investigation of Increasing the Cybersecurity of the Systems in Industry 4.0 using Artificial Intelligence” by the contract КП-06-Н47/7 with National Science Fund under the Ministry of Education and Science in Bulgaria.

Impact of selected teleconnection pattern on solar energy potential in different climatic zones of Nigeria

O F Awe[1,2], S T Ogunjo[3], A B Rabiu[1], O J Ajanaku[1,2], I D Ijila[1,2]

[1] Centre for Atmospheric Research, National Space Research & Development Agency, Kogi State University Campus, Anyigba, Nigeria
[2] Atmospheric & Space Weather Research Laboratory, Centre for Atmospheric Research, Osun State University, Osogbo, Nigeria
[3] Federal University of Technology Akure, P.M.B 704 Akure, Ondo State, Nigeria

Email: yomiawe@gmail.com

Abstract: Meeting the energy demands and sustainable development goals in Nigeria requires investigation of potentials of alternative energy sources and possible challenges to their reliability. In this study, we investigated the impact of four (4) teleconnection patterns on the solar energy potential within different climatic zones of Nigeria. Our results indicate that there are weak and insignificant correlation between the studied teleconnection patterns and solar energy potential on the long run. However, monthly analysis suggests significant correlation values between all the teleconnection patterns studied and solar energy production within all the climatic regions of the country. Therefore, it is important to consider the role of teleconnection pattern in energy planning and forecasting within the region.

1. Introduction

Teleconnections are spatial patterns in the atmosphere that link weather and climate anomalies over large distances across the globe [1]. Large scale oscillations in the Atlantic Ocean include Tropical Northern Atlantic (TNA), North Atlantic Oscillation (NAO), and Tropical Southern Atlantic (TSA). The teleconnection patterns in the Pacific oceans include El Nino Southern Oscillation (ENSO) and Pacific Decadal Oscillation (PDO). In the Indian Ocean, there is the Indian Ocean Dipole (IOD). A nonlinear relationship between different teleconnection patterns has been established [2]. It has long been recognized that the weather at one location on the Earth is linked to the weather at other very distant locations. ENSO has been found to affect drought in West Africa [3], locust outbreaks [4], fish population [5], and diseases in Africa [6]. However, the impact of teleconnection patterns on the energy sector has not been investigated.

There is an increasing demand for clean and affordable energy to mitigate the impact of climate change [7]. Solar energy is one of the most demanded renewable energies across the world. It has the advantages of being infinite and environmentally friendly. Recent technological advances have increased the efficiency of commercially available solar panels. Tropical countries, due to the higher insolation received, are adopting solar energy as the preferred renewable energy mode. However, there are several challenges to the deployment of solar energy, especially in tropical countries. These challenges include cloudiness [8], the initial cost of setting up, unclear national policy, and battery technology. Stoyanov et al [9] assessed the impact of such factors on the electricity generated from operational grid connected photovoltaic system (PVS). One of the unexplored challenges to solar energy in sub-Saharan Africa is the impact of teleconnection patterns.

Solar energy is one of the cleanest and most readily available renewable energies all over the globe, and for the sake of this research, we would be deliberating more on the impact of teleconnection patterns on solar energy potentials in the Coastal, Sahel and Guinea regions. The effect of solar activity variations on climate is far from being correctly quantified, which leads to difficulties in distinguishing between various causes of climatic variations [10, 11]. Research over the past century has shown that teleconnection patterns fluctuate over a very broad range of time scales, from just beyond the period of synoptic-scale variability, to inter-annual and even inter-decadal time scales. Besides the fact that teleconnections connect weather and climate over a great distance, which is a fascinating observation in its own right, teleconnections generate much interest because they persist for a longer time than most other phenomena in the atmosphere. This implies that when teleconnections are active, the atmosphere may be in a state that is more amenable to useful weather forecasts over a time period that is much longer than a typical weather forecast. The features that cause teleconnections to be long lasting are their long wavelength and their slow growth and decay.

Understanding and quantifying the contributions of teleconnection patterns to solar energy potentials in the tropical region will help to plan better for maximum power output. In this study, we aim to investigate the impact of teleconnection patterns within the Atlantic and Pacific oceans on the solar energy potential within different climatic zones at different temporal scales. Four teleconnection patterns in the Pacific and Atlantic oceans were considered across three climatic zones of Nigeria. Results from this study will help the country plan for the achievement of the Sustainable Development Goals (SDGs) through strategic planning towards accessing sustainable energy.

2. Methodology

2.1. Data and study area

For this work, the monthly 2m temperature and surface solar radiation data used for the Sahel, Guinea, and Coastal regions of Nigeria, were obtained between January 1980 to November 2020 from ECMWF Reanalysis 5th Generation data of the European Centre for Medium-Range Weather Forecasts [11]. While the teleconnection data used are the Tropical Southern Atlantic (TSA) Index, Tropical Northern Atlantic (TNA) index, Pacific Decadal Oscillation (PDO), and El Nino Southern Oscillation (ENSO). These data were obtained from the NOAA (National Oceanic and Atmospheric Administration) database [13]. The different climatic zones considered within Nigeria is shown in figure 1.

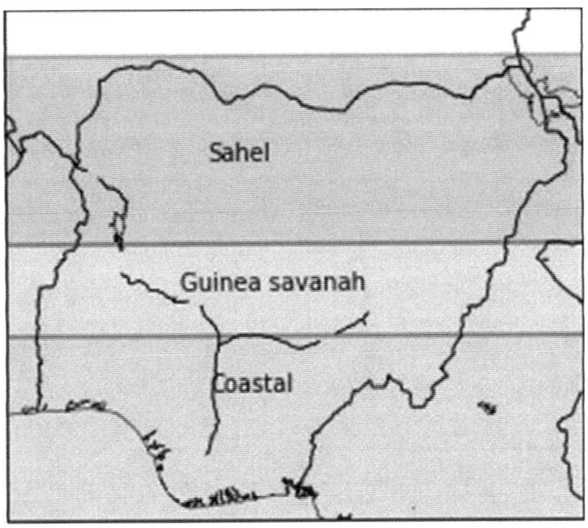

Figure 1. Classification of Nigeria considered in this study.

2.2. Solar energy potential

The energy potential is estimated using equation (1) [14, 15]

$$P = \eta_{T_{ref}} A G_T\left(1 - 0.0045(T_C - 298.15)\right) \quad (1)$$

where $\eta_{T_{ref}}, T_c, G_T, A$ are the cell electrical efficiency at reference temperature taken to be 0.14, air temperature in Kelvin, solar radiation flux on the module, and Area taken as 0.5 square metres respectively. The estimated solar potential for the three regions under consideration is shown in figure 2 while the temporal evolution of the teleconnection patterns is shown in figure 3.

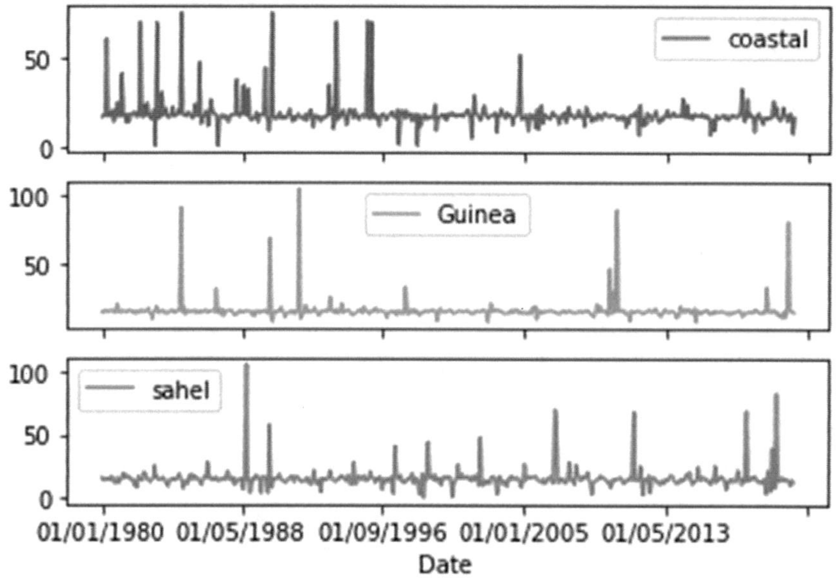

Figure 2. Monthly time series of estimated solar energy potential in the various regions.

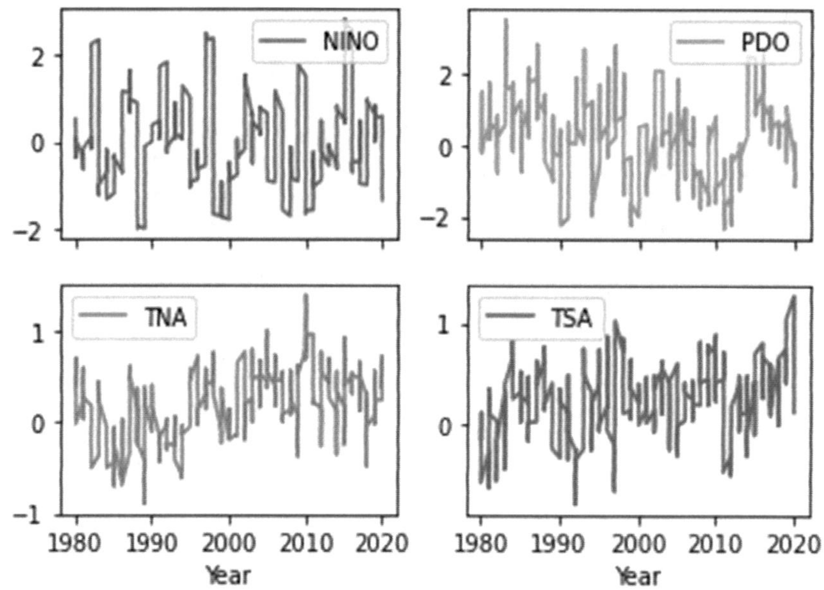

Figure 3. Time series of teleconnection patterns considered in this study.

3. Results and Discussion

The linear relationship between teleconnection patterns and the solar energy potential in the different climatic zones of Nigeria is shown in figure 4 while the statistics is presented in table 1.

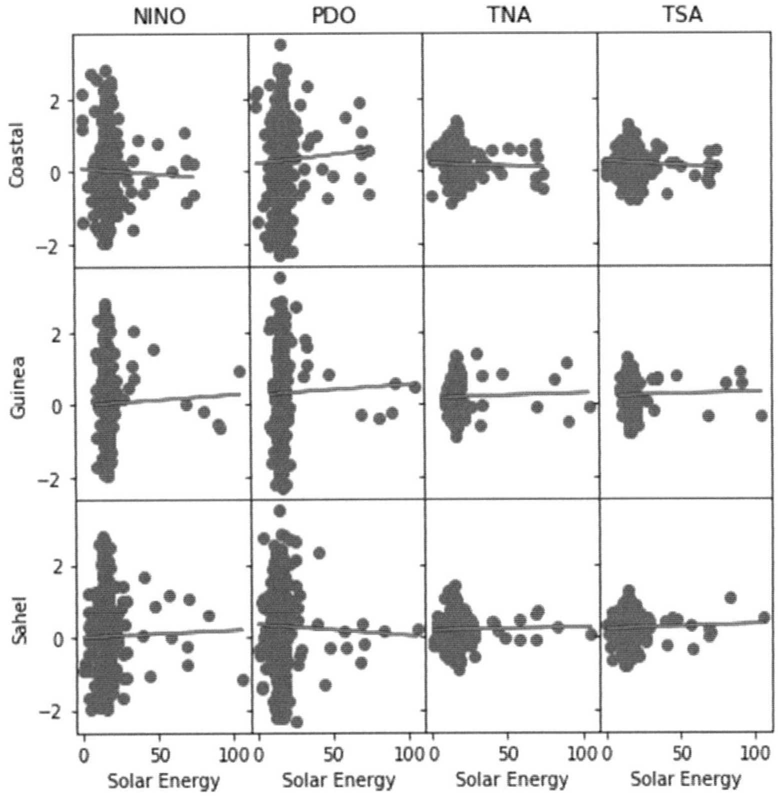

Figure 4. Correlation between teleconnection patterns and solar energy potential for data between 1980 – 2020

The correlations were found to be very weak and insignificant. In the coastal and Sahel regions, PDO showed negative and positive relationships respectively. All teleconnection patterns exhibit a positive correlation with the potential solar energy in all climatic region. The strongest correlation was observed between TSA and the solar energy potential in the coastal region while the weakest was with ENSO in the Sahel. Generally, ENSO and TSA showed a stronger correlation in the coastal region and weakest in the Sahel. Although, PDO and TNA exhibit the strongest correlation in the coastal region, their weakest correlation was in the Guinea region.

Table 1. Correlation statistics between teleconnection pattern and solar energy potentials in different climatic regions of Nigeria.

Region	NINO	PDO	TNA	TSA
Coastal	-0.029	0.036	-0.028	-0.055
Guinea	0.024	0.020	0.025	0.024
Sahel	0.017	-0.025	0.018	0.032

The monthly correlation values between solar energy potential and teleconnection pattern were studied and the results presented in figure 5. In the coastal region, strong negative significance was observed between NINO and PDO, and solar energy potential. The two then exhibit a positive relationship in April, although not significant. This implies that the increasing strength of NINO will likely reduce the solar energy potential in the coastal region in all months of the year except February, April, and November.

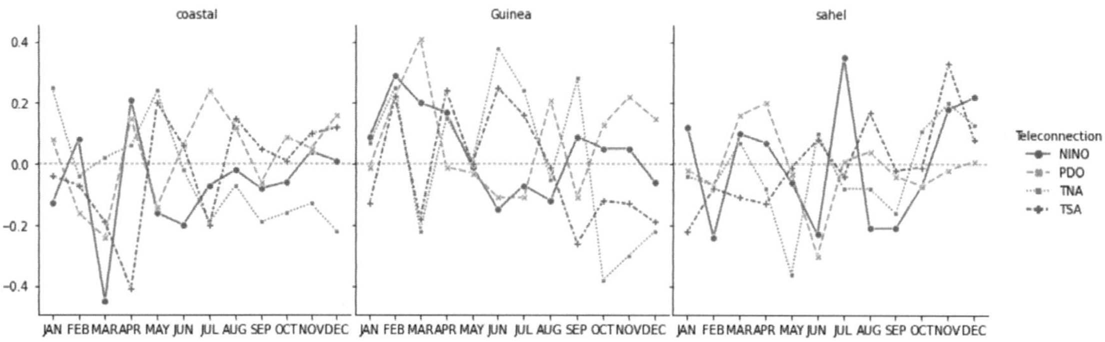

Figure 5. Monthly correlation between teleconnection pattern and solar energy potential in the different climatic zones of Nigeria

TSA showed a similar pattern of significant negative correlation in April and non-significant positive correlation in May. Furthermore, TNA showed strong positive correlation with solar energy potential in the coastal region in the months of January and May while negative correlations were observed for the teleconnection pattern in the later part of the year (June - December). The correlation of the teleconnection patterns with solar energy potentials was found to weaken towards the end of the year. In the Guinea region, significant correlation was reported for PDO (positive), TNA (positive), and TNA (negative) in March, June, and October respectively. TNA showed negative correlations in only four months of the year in the coastal region, but four months of positive correlation in the Guinea region. A strong significant correlation was reported in the Sahel region for TNA (negative), PDO (negative), NINO (positive), and TSA (positive) in the months of May, June, July, and November respectively.

4. Conclusion

In this study, the plausible impact of selected teleconnection patterns on the solar energy potential of three climatic regions within Nigeria was considered. Power production within the region; usually hydroelectric or thermal plants have suffered many setbacks including limited production, environmental challenges, disruption of the gas supply network, and expensive maintenance cost of the plants.

Our results suggest that at the annual scale, the teleconnection patterns have little or no impact on solar energy potential in Nigeria. However, monthly analysis suggests the significant impact of teleconnection on solar energy potential which must be considered during the planning of energy systems within the country. It is proposed that further research into the mechanism for these monthly correlation patterns should be investigated.

5. References

[1] Feldstein S B and Franzke C L 2017 Atmospheric teleconnection patterns *Nonlinear and stochastic climate dynamics* p 54

[2] Ogunjo S T and Fuwape I A 2020 Nonlinear characterization and interaction in teleconnection patterns *Advances in Space Research* **65** 2723–32

[3] Ogunjo S T, Fuwape I A and Olusegun C F 2019 Impact of large scale climate oscillation on drought West Africa *arXiv preprint arXiv:1901.10145*

[4] Todd M C, Washington R, Cheke R A and Kniveton D 2002 Brown locust outbreaks and climate variability in southern Africa *Journal of Applied Ecology* **39** 31–42

[5] Cheal A J, Delean S, Sweatman H and Thompson A A 2007 Spatial synchrony in coral reef fish populations and the influence of climate *Ecology* **88** 158–69
[6] Mabaso M L H, Kleinschmidt I, Sharp B and Smith T 2007 El Niño Southern Oscillation (ENSO) and annual malaria incidence in Southern Africa *Transactions of the Royal Society of Tropical Medicine and Hygiene* **101** 326–30
[7] Shahnazi R and Dehghan Shabani Z 2020 Do renewable energy production spillovers matter in the EU? *Renewable Energy* **150** 786–96
[8] Danso D K, Anquetin S, Diedhiou A and Adamou R 2020 Cloudiness Information Services for Solar Energy Management in West Africa *Atmosphere* **11** 857
[9] Stoyanov I, Ivanov V, Iliev T and Ivanov H 2020 Yield and Performance Study of a 1MWp Grid Connected Photovoltaic System in Bulgaria *Journal of Engineering Science and Technology Review* pp 206-209
[10] Gray L J, Beer J, Geller M, Haigh J D, Lockwood M, Matthes K, Cubasch U, Fleitmann D, Harrison and Hood L 2010 Solar influences on climate *Reviews of Geophysics* **48**
[11] Solomon S, Manning M, Marquis M and Qin D 2007 *Climate change 2007-the physical science basis: Working group I contribution to the fourth assessment report of the IPCC* vol 4
[12] Hersbach H, Bell B, Berrisford P, Hirahara S, Horányi A, Muñoz-Sabater J, Nicolas J, Peubey C, Radu R and Schepers D 2020 The ERA5 global reanalysis *Quarterly Journal of the Royal Meteorological Society* **146** 1999
[13] Anon 2020 National Oceanic and Atmospheric Administration (NOAA) *Monthly Atmospheric and Ocean Time Series*
[14] Ogunjo S T, Obafaye A A and Rabiu A B 2021 Solar energy potentials in different climatic zones of Nigeria *IOP Conference Series: Materials Science and Engineering* **1032** 012040
[15] Jie J, Hua Y, Gang P, Bin J and Wei H 2007 Study of PV-Trombe wall assisted with DC fan *Building and Environment* **42** 3529–39

Acknowledgements
This work is funded by the Centre for Atmospheric Research of the National Space Research and Development Agency, Nigeria, under the R & D fund.

Types of smart systems and their application in power grids

M E Mehmed-Hamza[1], A B Filipov[2] and M D Ivanova[2]

[1] Department of Medical Equipment, Electronic and Information Technologies in Health care, Medical University of Varna, 55 Marin Drinov Str., 9002 Varna, Bulgaria
[2] Department of Electric Power Engineering, Technical University of Varna, 1 Studentska Str., 9010 Varna, Bulgaria

E-mail: m.dicheva@tu-varna.bg

Abstract. With the development of modern technologies, opportunities for an exchange of the increasing amount of information at ever higher speeds are created. The use of smart systems in electric power engineering creates conditions for improving the systems for monitoring, archiving, analysis and management. It is necessary to collect the basic information about the emergency modes automatically for analysis of the operation of the equipment and the parameters of the regimes in the power grids. The main questions to be decided are: what type of information is needed, how to use this information and at what are the different access levels, what should be the systems - local or global, how to exchange data. Smart systems for power grids and switchgears have been developed on a modular basis by almost all global manufacturers, as well as by many companies on individual assignments. The publication discusses the existing types of smart systems and the data they process. Suggestions have been made for the volume and structure of data that are needed for the analysis process.

1. Introduction

Power distribution grids are considered for the purpose of the present study. When choosing the technical and economic solutions for the design and operation of facilities of the power distribution grid, it is necessary to observe mainly the requirements for:

1. security of electricity supply;
2. electricity quality;
3. economy;
4. the lowest possible operating and technological costs for electricity distribution;
5. easy and fast maintenance
6. safety of humans and animals;
7. impact within admissible limits on other technical systems;
8. protection of the environment, as well as aesthetic inclusion of the facilities in it.

It is necessary to collect the basic information about the modes and mostly about the emergency ones automatically for analysis of the operation of the equipment and the parameters of the regimes in power grids [1].

With the development of modern technologies, opportunities for an exchange of the increasing amount of information at ever higher speeds are created. The use of smart systems in electric power engineering creates conditions for improving the systems for monitoring, archiving, analysis and management [2].

It is important to note that due to the network structure of power distribution grids, data collection points are decentralized and in many cases the smart systems in the individual sites are different from each other. For example: data on electrical parameters are collected mainly in substations (SS); electrical, circuit and other data are systematized in dispatching centres, and meteorological data should be added from the territorial metrological stations and / or from local stations and systems with decentralized location near the respective power lines, SSs and other facilities.

The scope of the data is partially defined in [3], where the essential requirements and rules for conducting normal, economical and trouble-free operation of the equipment, buildings and facilities in the power plants and grids are determined.

The complex of technical means in automated control systems (ACS) includes:

- means for collecting and transmitting information;
- means for processing, visualization and archiving of information;
- management tools;
- auxiliary devices and systems.

By-laws and norms have been introduced in order to ensure the normal operation of each level in the electric power system (EPS). The rules for the operational information exchange and the information systems of the electricity system operator and electricity distribution grids shall comply with the similar rules in force at the Union for the Coordination of the Transmission of Electricity (UCTE) and the European Transmission System Operators (ETSO).

The information exchange should provide possibilities:

- for automatic processing of the operational interrelations between all parties;
- conclusions from the operating work and/or the emergency events, which shall be taken into account in subsequent corrective actions to be drawn;
- to facilitate the assessment of the possible risk that may arise and to determine appropriate actions for ensuring the reliability of operation and the integrity of the EPS.

The requirements for the details of the information must be defined in:

- the operative informing;
- the reports on events in the system;
- joint investigation of the events;
- registration and coordination of the information at commissioning and decommissioning of the facilities.

When exchanging information, the parties are obliged to:

- ensure transparency and accuracy of the information;
- observe the necessary confidentiality when their market interests are affected.

The information to be exchanged must be sufficiently detailed to describe the order or the operational action, to enable the recipient of the message to comply with it and to assess the possible risk. The information must include the name of the person reporting the operational work or the order on behalf of the user or the electricity system operator. The recipient may ask questions to clarify the information and the informing party must ensure that the necessary information is provided.

With the development of Smart Grids, it is essential that information on different events in power grids to be properly recorded and systematized in the data arrays for their subsequent analysis (for example for outage cause detection) using different algorithms [4, 5].

The purpose of the research is to systematize specific uniform logs based on analysis of the activities of the electricity distribution and transmission companies. The publication discusses the existing types of smart systems and the data they process. Creating a list of standardized logs to describe the types of outages will allow for easier and more accurately analyzing the outages. Proposals have been made for the volume and structure of data that are needed for the analysis process.

2. Analysis and discussion

A summary of some of the available smart systems by manufacturers is presented. For each system, table 1 provides information on the type of communication and the protocols for data exchange [6 - 10].

Table 1. Communication and protocols for data exchange.

System	Communication	Protocol
SIEMENS SICAM PAS Power Automation system	Local Ethernet Connection	IEC 60870-5-101, IEC 60870-5-104, IEC 61850, IEC 60870-5-101, IEC 60870-5-103, PROFIBUS FMS
ABB SMS 510 Substation Monitoring	Local or remote through network. TCP/IP	
ABB MicroSCADA	Versatile access to mobile phones (WAP, SMS, GPRS) Local HMI control	IEC 61850, IEC 60870-5-101, IEC 60870-5-103, IEC 60870-5-104.
GENERAL ELECTRIC Substation Automation System	Wonderware InTouch™ or GE Fanuc's Cimplicity HMIs	RS 232 and RS 485, fibre - optic connections in LAN or WAN according to the standard TCP / IP. GE-SA
ROCON SCADA-system MMI D (Man Machine Interface) RSS 200	local and remote IEC / ANSI / EN	IEC 61850 IEC 61850, IEC 60870, MODBUS, DNP, etc.

All three companies (operators) of the transmission or distribution power grids in Bulgaria and their operational duty personnel must fill in round-the-clock statements in established forms.

These forms are normatively defined, from which the filling of the content should be of the same type. However, the diversity of ACS in different companies and the common technical capabilities cause differences in the form content.

All companies work on an operational executive scheme of the transmission / distribution power grid (layout scheme) in which the scheme elements are of the same type (circuit breaker, power transformer, etc.) but with different degree of detail and symbolic designations.

All of the commissioned ACS in the electricity distribution companies and in ESO collect, process and archive data and present them in generally accepted forms as follows:

- operational diary;
- logbook for the request for decommissioning of the transmission / distribution power grid facilities;
- logbook for the relay protections and automation;

- logbook for the operation of relay protections;
- diary of order assignments;
- logbook for the equipment damages;
- logbook for the trips of the circuit breakers;
- map for the settings of the relay protections and automation;
- others.

In all companies the form of the paper operational diary is according to the normatively established form, but the used ACS are different. For example, a map for the settings of relay protections and automation used in three of the companies has the form presented in tables 2, 3 and 4.

Table 2. Map of the settings of relay protections and automation in electricity distribution company 1.

Cell №	**Transformer/ Line**	**Instantaneous Protection (IP)**				**Time overcurrent protection (TOCP), Step I**				**Auto-reclosing (AR)**	Note
		Type	*I*I_n*, A*	*t, s*	*tional (Yes/*	*Type*	*I_n*	*t*	*Directional*	*Type*	
			A	s			A	s	Yes/No		
1	XXX	7SR224	1000	0	No	7SR224	300	0.2	No	Yes	AR: time setting 1s, activated from TOCP with blocking from IP

Table 3. Map of the settings of relay protections and automation in electricity distribution company 2.

№	**Line**	**Current transformers**	**Type of protection**	**Primary current setting**	**Time**	**Type of protection**	**Current setting**	**Time**
1	XXX	200/5/5 (dry type) - 2	Earth fault protection (EFP)	80 A	0 s	Time overcurrent protection (TOCP)	280 A	1 s

Table 4. Map of the settings of relay protections and automation in an electricity transmission company.

Substation	**Power line**	**Zone I** Ω	**t** s	**Zone II** Ω	**t** s	**Zone III** Ω	**t** s	**Zone IV** Ω	**t** s
XX	XX	3.15	0.0	15.06	0.7	23.33	0.9	-	5.9
		Starting relays Q4				**Control points of R-X Diagram**			
		I, A	I, A			$U_{pick-up}$,V	I, A	$U_{drop-out}$,V	I, A
		$A_i = 0.4 \cdot I_n$	A_e – 50 %, $1.5 \cdot I_n$			46.0	5.0 A and 66°	48.0	4.0 A and 20°30’
		EFP, step I		**EFP, step II**		**Auto-reclosing with synchronism and voltage absence check**			*Note*
		$I_{primary}$, A	t, s	$I_{primary}$, A	t, s	U, V	t, s	δ, °	
		1	XXX	6400	0.6	40	0.8	33	

The differences are obvious, despite the fact that the information presented is informative and understandable. The situation is also identical with the filling of an operational diary. Main differences

IOP Conf. Series: Materials Science and Engineering 1216 (2022) 012003 doi:10.1088/1757-899X/1216/1/012003

are obtained when filling out data for "operations during work shift". When comparing and analyzing the logs in the other diaries, similar filling by the three companies is also observed.

After processing the data from the ACS, the results for number and duration of power outages by type of activities/causes for the electricity distribution companies 1 and 2 are systematized in tables 5 and 6. For the first company the causes are 26, and the for the second one – over 100.

Table 5. Outages by type of activities for electricity distribution company 1.

Causes	2018	
	count	**min**
Emergency and restoration activities	68	4923
Restoration of the normal scheme	62	2082
Lightning	27	847
Animals	11	520
Amendment of the normal scheme	2	45
Investments	190	194778
Abnormal weather conditions - wind, torrential rains, hail, ice	1180	76891
Thefts and other criminal acts	2	239
Uncoordinated excavation, construction and other activities of third parties	84	4700
Maintenance	647	206493
Overhaul	96	79947
Overhaul/Investments	56	166826
Power failure from HV grid	2	3
Planned one-off measure	454	102894
Fire	13	1971
Fire, industrial accidents	1	15
Power supply disconnection of a consumer	19	413457
Activities of employees of the electricity distribution company	6	77
Activities of employees of the Natsionalna Elektricheska Kompania EAD/ Electricity System Operator EAD	2	24
Activities of employees of third parties	14	14399
Natural disasters - floods, earthquakes, drought, fires, landslides, etc.	3	645
Birds	31	1558
Facilities owned by third parties	456	183497
Facilities of the electricity distribution company	3499	267262
Medium voltage facilities of the Natsionalna Elektricheska Kompania EAD/ Electricity System Operator EAD	26	807
Cable mechanical damage	8	786
Total	**6959**	**1725686**

Table 6. Outages by type of activities for electricity distribution company 2.

Causes	2018	
	count	**min**
Construction and assembly works (CAW)		
Excavation works on the cable route		
Work order assignment №XX – replacement of indirect-measuring power meter		
Damaged cable line from Transformer station №XX to Distribution box №XX at CAW		
Maintenance of the Main Switchboard (MSB) of hotel " XXX"		
Self-disconnection of an automatic circuit breaker in MSB1 in hotel "XXX"		
Restoration of the normal scheme of an overhead power line of Transformer station №XX		
Disconnected overhead power line in Transformer station №XX – time overcurrent protection		
Disconnected line "XXX" – instantaneous protection		

More over 100 causes		
Total	**152**	**10060**

For the second company, a summary of the results for reducing the causes could be made - table 7.

Table 7. Summary of the outages by type of activities for electricity distribution company 2.

Causes	**2018**	
	count	**min**
In Transformer stations	66	7622
Due to a reason determined by the customer	13	1563
In Distribution boxes	8	355
In or after a Distribution box	1	20
By cable line	7	195
Mast transformer station	1	305
Distribution box		
In substations		
Other	56	0
Total	**152**	**10060**

After analyzing the described causes, it can be concluded the following:

- The layout of the records is dissimilar in the different companies.
- In electricity distribution company 1, a foreseeable number of causes are defined, a total of 26.
- In electricity distribution company 2, the causes are not systematized and the records are with individual layouts.
- It is allowed to mix information about the facility, the place (location or facility in different logs) and the cause that led to outages.

3. Conclusion

Based on the analyses made, in order to increase the efficiency of the automated control systems and the results of each company in terms of power quality of consumers, we propose that the companies introduce changes in the automated control systems, namely:

1. In the automated control systems to systematize information separately about:

- **Facility type** - For example: Substation, Power line, Transformer station, Distribution box, Building electrical installation or the respective abbreviations for short record.
- **Facility property** - For example: Owned by the company, Foreign property – of an electricity transmission company, Foreign property – customer (user), Foreign property - customer (manufacturer).
- **Disconnection type** - For example: Automatic, Manual
- **Cause 1** - For example: Planned, unplanned.
- **Cause 2** - For example: Table 8.

Table 8. Proposals for type description of logs for outages by types of activities.

	Causes by types of activities
1	Planned maintenance
2	Unplanned service/repair
3	Activity of the company
4	Activity of an electricity transmission company
5	Third party activity
6	Amendment of the normal scheme
7	Restoration of the normal scheme
8	Emergency and restoration activities
9	Atmospheric causes, values below the designed ones for the facilities
10	Atmospheric causes, values above the designed ones for the facilities
11	Natural disasters - floods, earthquakes, droughts, fires, landslides, etc. National or regional

12	Thefts and other criminal acts
13	Animals
14	Other

2. The input of the data for each event to be regulated normatively in respect of standardized structural logs.

The above changes could be introduced by increasing the opportunities for analysis of the technical condition of facilities, the causes of power supply interruptions and more.

4. References

[1] Ivanova M, Dimitrova R and Filipov A 2020 Analysis of power outages and human errors in the operation of equipment in power grids *XIIth Electrical Engineering Faculty Conference, BulEF 2020* (Varna, Bulgaria) DOI: 10.1109/BulEF51036.2020.9326058

[2] Borlase S 2017 *Smart Grids:Infrastructure, Technology, and Solutions* (CRC Press)

[3] Ordinance № 9/ 9.06.2004 for the technical operation of power plants and networks. 19.12.2004, (In Bulgarian), Accessed April 2021.

[4] Zhang Y, Huang T and Bompard E 2018 Big data analytics in smart grids: a review *Energy Informatics* vol 1 Article number: **8**

[5] Bashkari S, Sami A and Rastegar M 2021 Outage Cause Detection in Power Distribution Systems Based on Data Mining *IEEE Transactions on Industrial Informatics* vol 17 iss 1 pp 640-649 DOI: 10.1109/TII.2020.2966505g

[6] https://new.siemens.com/global/en/products/energy/energy-automation-and-smart-grid/substation-automation/substation-automation-system-sicam-pas.html (Accessed April 2021)

[7] https://new.abb.com/medium-voltage/distribution-automation/protection-relay-services/legacy-relays-and-related-devices-and-tools/substation-monitoring-system-sms-510 (Accessed April 2021)

[8] https://www.gegridsolutions.com/automation.htm (Accessed April 2021)

[9] http://www.roconbg.com/bg/engineering-services/sistemi-za-avtomatizaciq-3 (Accessed April 2021)

[10] Stoyanov I, Iliev T, Mihaylov G, Ivanova E and Kogias P 2018 Smart Grid Communication Protocols in Intelligent Service for Household Energy Use *Advances in Intelligent Systems and Computing* 661 pp 380–389

Assessment of energy efficiency of a motor yacht depending on routes and sailing area

G. Ivanova [1], V. Gyurov [2]

[1,2] Department of Electric Power Supply and Electrical Equipment , Technical University of Varna, Bulgaria

E-mail: ginkahivanova@tu-varna.bg

Abstract. The report presents an analysis of luxury charter yachts which are a specific class of vessels that, in order to meet the schedules and increased requirements of the owners regarding the luxury of the services offered, have relatively high energy consumption for their gross tonnage, compared to other passenger ships. The study focuses on the analysis of energy efficiency of luxury yachts by calculating the energy efficiency index (EEDI). This involves comparing different parameters that affect the value of the EEDI and can lead to energy savings. The report presents theoretical and experimental studies of the energy costs of a 70-meter luxury yacht for charter trips. With the use of the design data of the ship's electrical equipment at different operating modes, at different routes, under different conditions, dependences for analysis of the optimal modes of movement are obtained, which leads to the realization of maximum savings of electricity and primary energy.

1. Introduction

Powerful motor yachts are designed for pleasure, they carry very few passengers and their energy efficiency requirements have almost no profile. A novelty on the cruise market, are the ultra-luxury expedition motor yachts. This relatively new type of ship is entering the market at a rapid pace of development. As the number of willing passengers continues to increase, operators seek to pre-determine their routes, serving selected interests and customer groups. Luxury motor mega yachts use modern technology to create a unique and exclusive trip. In an effort to reduce the global carbon emission of the maritime industry, the International Maritime Organization (IMO) has adopted rules to reduce greenhouse gases in terms of the efficiency of the offshore power plant. The strictest requirements are for emission control zones (ECA). Namely:

- Baltic and North Sea
- Coastal waters of the United States and Canada
- Caribbean Sea
- Mediterranean Sea
- Coast of Japan
- The Strait of Malacca, etc.

2. State of problem

The study on the implementation of EEDI in the yachting industry must be examined in detail, taking into account the energy efficiency requirements that are mandatory for all ships with a gross tonnage of more than 400 GT built after 1 January 2013. The equation that applies to EEDI is directly related to the benefits of operating a particular ship or its transport activities. Luxury motor yachts are designed for pleasure, they carry very few passengers, the calculations on the useful work or the number of passengers will not meet the requirements imposed by the IMO. It is therefore more difficult to determine any direct benefit or work done than other cruise ships, as the carrying capacity of cargo and passengers per nautical mile is very low. The final version of the achieved EEDI equation is provided by the IMO MEPC 62/WP. 15, ANNEX 1.

$$AttEEDI = (MEE + AEE - ERIT) \cdot \frac{1}{TW} \tag{1}$$

Where: MEE - CO_2 emissions from main engines [tCO_2.DWT]; AEE - CO_2 emissions caused by auxiliary engines [tCO_2.DWT]; ERIT - emission reductions from innovative technologies [tCO_2.DWT]; TW - transport work [DWT];

The calculation of MEE, AEE, ERIT and TW is performed by formulas 2 and 3.

Calculation of MEE - emissions from the main engines:

$$MEE = \left(\prod_{j=1}^{n} f_j\right) \cdot \left(\sum_{i=1}^{n_{ME}} P_{ME(i)} \cdot C_{FME(i)} \cdot SFC_{ME(i)}\right) \tag{2}$$

Where: PME (i) [kW] - 75% of the maximum continuous load (MCR) of the main engines.

$$P_{ME(i)} = 0.75 \cdot (MCR_i - P_{PTO}) \tag{3}$$

Where: P_{PTO} [kW] - 75% of the nominal mechanical power of the shaft generator divided by efficiency on the shaft generator; PME: Main engine power (kW), $P_{ME\ (i)}$ is 75% of the total rated installed power (MCR) after deducting shaft generators, of course. $CF_{ME\ (i)}$ [t-CO_2 / t-Fuel] - environmental impact factor for fuels; $SFC_{ME\ (i)}$ [t-Fuel / kWh] - specific fuel consumption of main engines at 75% MCR load; f_i - correction factor taking into account the influence of specially designed energy efficient elements of the ship.

Instead, the IMO guidelines advise the use of 1 if there are none, as is the case with luxury yachts. Usually this correction factor is very important as it takes into account the ratio of L_{BP} (Length between perpendiculars) and installed power (MCR). Reducing the average weight of the installed power on board a luxury yacht will result in a correction of this value, which will be less than 1.

As seen in the above EEDI equation, there are different correction factors and constants. Some of these coefficients will have a direct impact and they are also applicable to yachts.

IMO provides simplified formulas that can be used, depending on the installed capacity. Unless different from those shown below:

If:

$$\sum MCR_{ME(i)} > 1000kW, P_{AE} = 0.025 \times (MCR_{ME} + 250) \tag{4}$$

and if:

$$\sum MCR_{ME(i)} < 1000kW, P_{AE} = 0.05 \times MCR_{ME} \tag{5}$$

Where: SFC - Specific fuel consumption [g/kWh] is provided by the engine manufacturers on board the yacht. In the case of yachts, ordinarily marine diesel and gas oil are used, as results carbon content equal to 87.44% and C_f measure in t-CO_2/t-Fuel equals to 3.206 (according to IMO MEPC EEDI circular and ISO 8217).

3. Proposals for assessing the energy efficiency of a motor yacht according to the different sailing regimes in different areas

We will do an experimental study of the relationship between the optimal speed of a luxury motor yacht, the route, energy costs and EEOI (Energy Efficiency Indicator), which means "optimal yacht speed", which improves operational energy efficiency [1, 2, 3]. The yacht with traditional propulsion, with a length of 70.54 m, class LR X 100 A1 SSC, G6, XLMC, UMS is considered. Main engines: 2xCaterpillar 3516B (2x1825kW @ 1800 rpm), generators: 3 x 200Kw, max. speed 17Kn, total power 4894 hp. The capacity of the fuel oil tanks is 204,000 l. This study will analyze the design data of a motor yacht and accordingly will propose an idea for improving energy efficiency in relation to different sailing areas and IMO requirements in these areas. Improving the efficiency of a power plant essentially involves burning less fuel and thus releasing less carbon into the atmosphere. The most important function of a yacht is to provide a high-quality form of recreation for guests on board. How well each ship performs in this function is reflected in the "luxury" goal. The speed range is 13 knots for cruising speed and 17 knots for maximum speed. With the Greenhouse Gas Strategy 2050, the IMO aims to reduce carbon intensity by 40% over the next decade by 2030 and by a total of 50% (intensity 70%) by 2050 [4, 5]. When calculating the cost of yacht fuel, you must first understand how far the yacht will travel. On average, marine diesel engines burn about 0.4 kilograms of fuel per hour per unit of horsepower. Another important factor is the meteorological conditions in the area in which the yacht is chartered, such as the condition of the water, wind, resistance and weight of the vessel, which affect the amount of fuel consumed. After mapping the route and calculating the distance in nautical miles in combination of fuel consumption per hour and the expected travel time, we can calculate our fuel costs on the yacht for a given route in a certain area. The ratio of GT to installed power (kW) is an important aspect for shipbuilders, as it can provide them with a quick and easy method for predicting whether a yacht is EEDI efficient [6].

We can calculate the average amount of CO_2 emissions from the main engines of the yacht (tonne of CO_2), it can be calculated by multiplying the average daily fuel consumption of the yacht by the CO_2 emission factor of marine fuels (tonne CO_2/tonne of fuel). The analysis is based on the following assumptions:

- The yacht uses MGO as fuel when chartered to SECA, and HFO when chartered outside SECA (main engines). Auxiliary engines use MGO throughout the route.
- We assume that the fuel consumption function is cubic (the daily fuel consumption is proportional to the power of the sailing speed) and that the consumption does not vary depending on the fuel used.
- The yacht is provided for weekly routes.
- The route (series of preferred ports) is not fixed.
- Revenues for the service are fixed.

Energy efficiency is a measure of CO_2 emissions. CO_2 emissions are measured in kg/h at each yacht mode. The most important thing is to determine the area of navigation. The Mediterranean, the Caribbean and the Bahamas are the most popular destinations. However, there are destinations that are more preferred. For the analysis considered the most popular destinations, such as the Mediterranean for the summer season and the Caribbean for the winter season. The data considered can be found in table 1, which represents the maximum, minimum and average percentage of time for each mode of operation of the yacht.

Table 1. Yacht operating modes based on maritime trafic

Type	Cruising [%]	Crossing [%]	Max. Speed [%]	Manoeuvring [%]	At Anchor [%]	In Port [%]
Maximum	22.92	18.58	1.12	3.36	67	78.04
Minimum	1.8	0	0.14	0.14	10.21	6.23
Average	8.4	8.03	1.3	1.3	38.97	42.66

According to previous calculations, maintenance costs are measured in relation to the output power of these components, the maintenance of the engines and generators installed in the yacht, and are presented in table 2.

Table 2. Maintenance costs of the main diesel engines and generator sets for the given yacht

Type	Number Installed	Power Output [kW]	Maintenance [usd/hour]	Operational hour per year
Main Engine	2	1825	5.11	1591
Generator Set	3	200	3.05	5006

It is known that yachts often spend their time away from their home ports, either in anchorages or in other ports. First, the amount of CO_2 emissions per hour is calculated for each mode of operation of the yacht. The hourly emissions for the different operating modes are then combined with the operating profile, thus relating to the annual operating hours of the set mode. In this way, the total annual CO_2 emissions are quantified.

The percentage of the different operating modes in relation to the total working period are shown in figure 1.

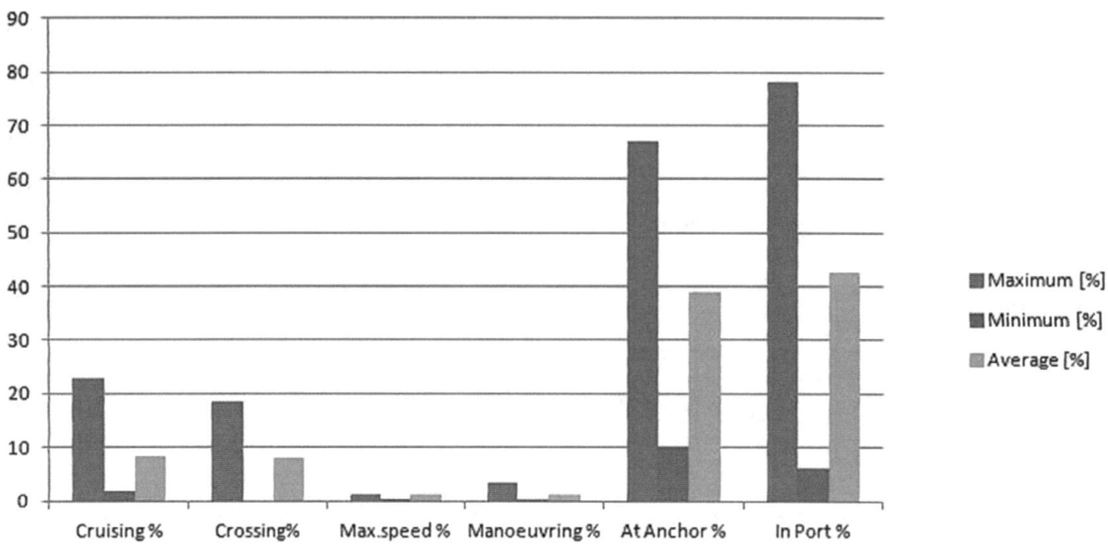

Figure 1. Yacht operating modes based on maritime traffic

A characteristic feature of luxury yachts is the significantly longer period in port or at anchor compared to the periods of sailing. The maintenance costs of the main diesel engines and generator sets for the given yacht are shown in figure 2.

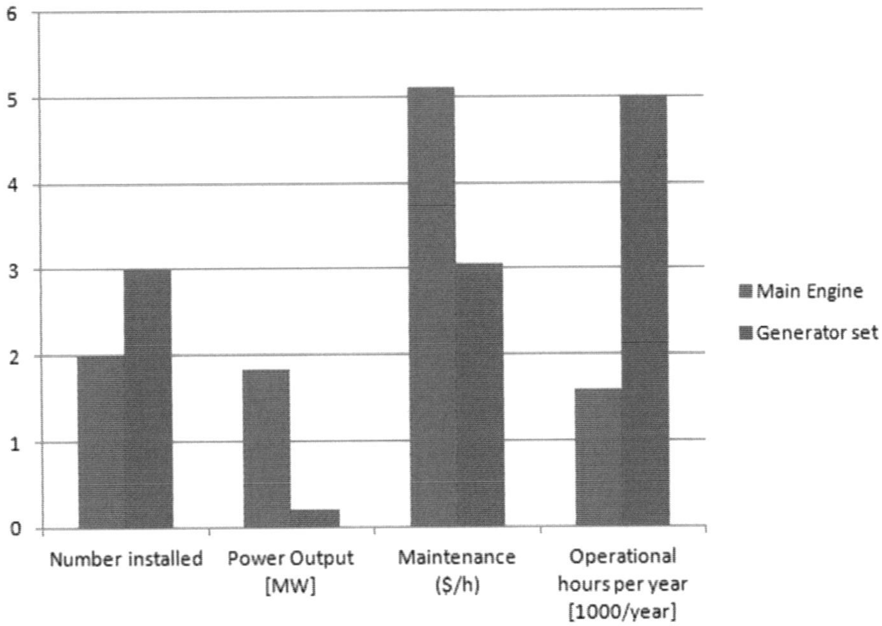

Figure 2. Maintenance costs of the main diesel engines and generator sets for the given yacht

An essential feature is the significantly higher hourly usability of the generators compared to the main engines. This is determined by the longer stays in port.

4. Conclusion
The report examines the specifics of energy consumption of luxury yachts for charter trips. The definitions for Energy Efficiency Design Index (EEDI) and their applicability for luxury yachts are considered. The results show quantitative dependences on the ratios between the different operating modes, energy consumption and maintenance costs. The significant difference between duration of regimes "sailing-harbour" shows the need for in-depth study of issues related to energy efficiency and environmental impact of yachts for various operating modes.

5. References
[1] Acomi N and Acomi A 2014 *Improving the Voyage Energy Efficiency by Using EEOI (Procedia - Social and Behavioral Sciences vol 138)* (Beijing: Elsevier) pp 531-536

[2] Kim S, Roh M, Oh M, Park S and Kim I 2020 *Estimation of Ship Operational Efficiency from AIS Data Using Big Data Technology (International Journal of Naval Architecture and Ocean Engineering* vol 12*)* (Seoul: Elsevier) pp 440-454

[3] Katarzina P and Igielski K 2018 *Calculation of Operational Indicator EEOI for Ships Designed to Other Purpose than Transport Based on a Research – Training Vessel (New Trends in Production Engineering* vol 1*)* (Warsaw: Sciendo) issue 1 pp 335-340

[4] Ren H, Ding Y and Sui C 2019 *Influence of EEDI (Energy Efficiency Design Index) on Ship–Engine–Propeller Matching (Journal of Marine Science and Engineering* vol 7*)* (Basel: MDPI) pp 1-22

[5] Sun C, Wang H, Liu C and Zhao Y 2020 *Real Time Energy Efficiency Operational Indicator (EEOI): Simulation Research from the Perspective of Life Cycle Assessment (Journal of Physics: Conference Series* vol 1626) (IOP Science) pp1-8

[6] Sun C, Wang H, Liu C and Zhao Y 2020 *Real Time Energy Efficiency Operational Indicator (EEOI): Simulation Research from the Perspective of Life Cycle Assessment (Journal of Physics: Conference Series* vol 1626) (IOP Science) pp1-8

Acknowledgments
The project discussed herein is supported by the European Regional Development Fund through the Operational Programme "Science and Education for Smart Growth" under contract UNITe № BG05M2OP001-1.001-0008. Project name: "National center of mechatronics and clean technologies".

IOP Conf. Series: Materials Science and Engineering 1216 (2022) 012005 doi:10.1088/1757-899X/1216/1/012005

Study on technical solutions for shore power supply of motor yacht

V. Gyurov[1], G. Ivanova[1]

[1] Department of Electric Power Supply and Electrical Equipment, Technical University of Varna, Varna, Republic of Bulgaria

E-mail: valentin.giurov@tu-varna.bg

Abstract. The report presents an analysis of modern technical solutions for shore power supply of a specific class of passenger ships - luxury yachts for charter trips. The design data of a motor yacht in its different operating modes are considered, as well as the energy mix related to the fuel consumption at shore supply of the yacht during its stay at the port. The use of modern technical systems for shore supply includes the application of specific frequency converters, through which compatibility between the different voltage standards and the frequency for different shore power supply systems is realized. The costs of onshore power, compared to those of marine fuel, can be calculated from the current prices of onshore electricity and the energy produced from its own generators. The analysis of the basic design data of the motor yacht and the assessment of the energy costs on board, fuel economy and emission reductions will provide a clear answer to the advantages of the power supply from the shore of the vessel.

1. Introduction

Motor yachts are a specific class of passenger ships, which in recent years have played a growing role in the overall energy consumption of shipping. The reason for this is the high demand for yachts that have more tourist than transport function and have a large sailing distance. Modern luxury yachts reach lengths up to 90 m, tonnage up to 2500 GT, speed up to 20 kn and passengers and crew up to 50 people. These parameters place the yachts in the class of small passenger ships, but with increased requirements for the comfort of the environment. For this reason, it is necessary to use powerful propulsion systems in the range of 2-4 MW and machinery equipment with high energy consumption in the range of 0.5-1MW. Luxury yachts also have significant specifics in terms of sailing regimes.

They are characterized by a much longer stay in ports than other types of ships. In some cases, this period can vary from days to several weeks. For this reason, in order to reduce fuel consumption and reduce greenhouse gas emissions, a shore power supply is used for all equipment and energy consumers on board. The use of shore power is related to the following problem - yachts can visit ports that have different standards for the level and frequency of the supply voltage (50 or 60Hz). This predetermines the need for analysis of the specifics, technical means and modes of electricity consumption of superyachts in shore supply mode. The problem is relevant, because different technical solutions for shore electric power supply have different advantages and disadvantages. The report presents a study of some types of technical solutions for shore supply of superyachts.

IOP Conf. Series: Materials Science and Engineering 1216 (2022) 012005 doi:10.1088/1757-899X/1216/1/012005

2. Study on the technical solutions for providing shore electric power supply for motor yachts

The basic standards for the on-board power supply system of yachts are: European – 400/230V 50Hz; USA – 460/265V, 240/120V 60Hz; China 380/22V 50Hz; Japan 220/110V 60Hz. This shows the existence of significant differences in voltage levels and frequencies for different countries. The safe connection of a yacht to the shore supply requires the application of special technical equipment.

2.1. Technical solutions for shore electric power supply with frequency conversion

Classical means for converting voltage and frequency include the use of specialized motor-generator groups (figure 1). In this case, the frequency conversion is performed by changing the speed of the electric motor. This can be done with specially designed electric motors or with the use of standard frequency regulators, allowing achievement of rotation above the synchronous speed. A significant disadvantage of the method is the weak unification - it is possible one-way conversion of voltage and frequency from one standard to another, but not in the opposite direction. The large mass and size parameters of the equipment do not allow to be installed on board a ship. Yachts may use this type of shore supply only in ports that are specially equipped with such facilities. Usually, this type of equipment is designed as mobile in standardized containers and can be moved from one port to another, but this involves significant additional economic costs, which significantly reduces the effect of their use. In practice, they are used where it is planned to permanently dock yachts which are produced according to different voltage standards from the shore network. Due to the double conversion of electricity, which is accompanied by additional power losses, the solution is less efficient than modern solutions without the use of electrical machines [1, 2].

Another way to convert voltage and frequency for shore supply is associated with the use of specialized galvanically separated inverters (figure 2). For this purpose, it is necessary to use a isolating transformer with sufficient power to transfer the required electricity. The advantage of the method is that due to the absence of electrical machines, greater energy efficiency and lower additional losses can be achieved. The technical means allows the two-way conversion of voltage and frequency - a yacht manufactured according to European standards to be powered by an American network or vice versa [3]. This type of equipment has smaller dimensions and can be produced both as additional equipment to the port infrastructure and to be manufactured for installation on board the yacht. Due to its great unification, this type of solution allows a yacht manufactured to a certain standard to be powered by all types of port infrastructure around the world.

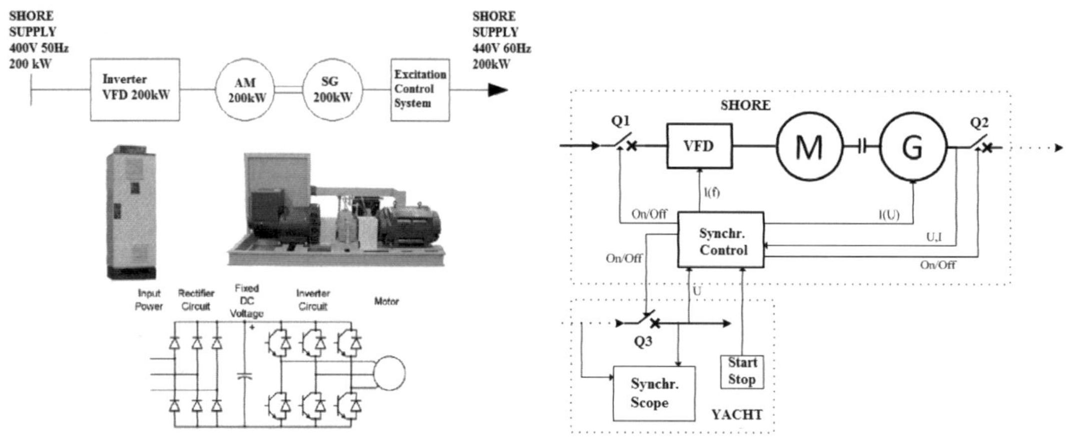

Figure 1. Voltage-frequency conversion system with M-G group for shore power supply

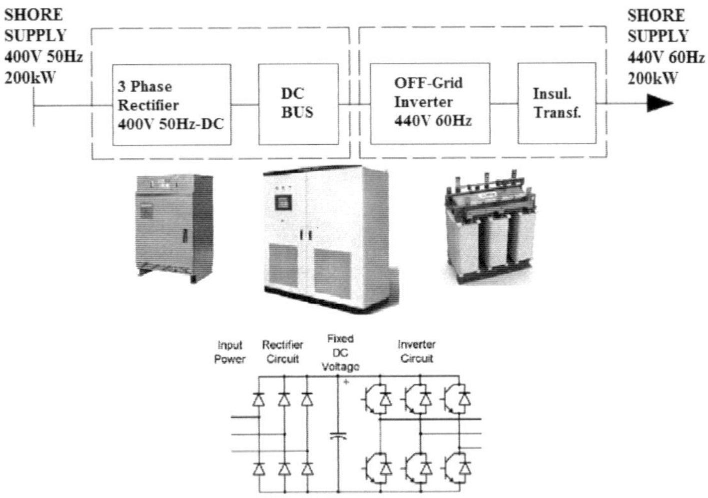

Figure 2. Voltage-frequency conversion with insulated inverter

2.2. Technical solutions for shore electric power supply without frequency conversion

This type of solution involves the use of a specialized multi-stage isolation transformer (figure 3). The initial connection of the windings is carried out with current-limiting resistors, as the stages are determined manually or with the use of a programmable logic controller. This solution is widespread for yachts whose electrical equipment is not sensitive to changes in feed frequency. The solution is applicable to older generations of yachts, where the main lighting is with incandescent light sources, and the electrical equipment is designed to operate at different frequencies. With the introduction of power electronics and frequency converters on board yachts, this solution is gaining popularity again.

This is possible if all electric drives on the yacht are equipped with inverters that can operate at different frequencies and all lighting systems have electronic drivers for LEDs or gas discharge sources operating on both basic standards.

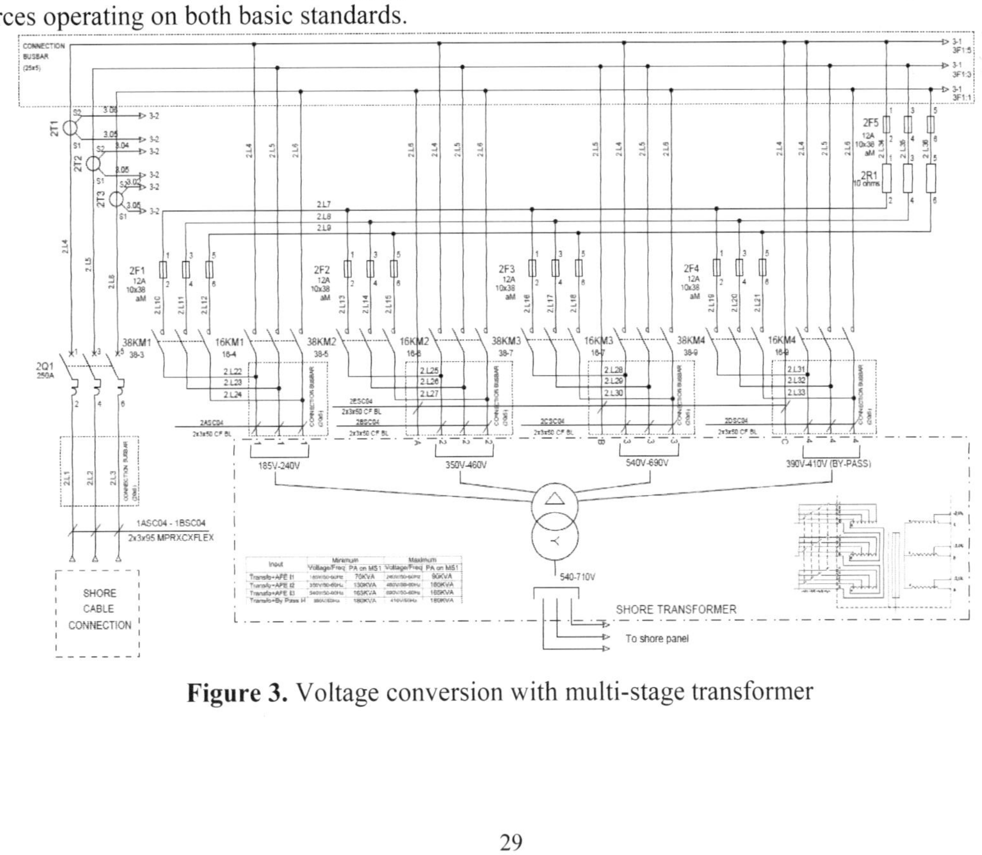

Figure 3. Voltage conversion with multi-stage transformer

3. Analysis of the power consumption of a motor yacht in shore power supply mode

3.1. Object of the study

The object of the study is a luxury yacht for charter trips with the following parameters: Length – L_{LW} = 70.5 [m], Gross Tonnage – GT=1560 [t], Max. speed – V_{MAX} = 17 [kn], Cruise speed – V_{CR} = 15 [kn], Crew = 39. A view of this class of yachts is shown in figure 4.

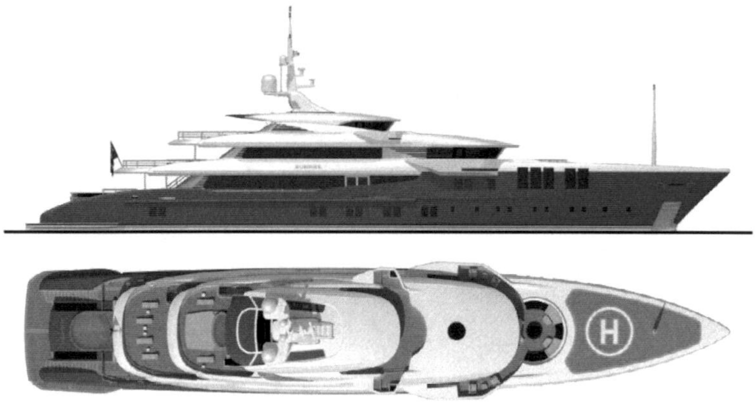

Figure 4. Superyacht GT 1560t, L=70.5m, Crew 39

3.2. Energy parameters of superyacht GT1560

The yacht is powered by two diesel engines 2x1825 kW and three diesel-electric generators 3x200 kW. The electrical consumers are divided into four main groups: Group 1 "Machinery Equipment" ($P_{INSTALLED}$ = 250.2 kW), Group 2 "Boiler/Fan/AC Unit and Water Supply" ($P_{INSTALLED}$ = 412.3 kW), Group 3 "Winches" ($P_{INSTALLED}$ = 82.6 kW) and Group 4 "Other Equipment" ($P_{INSTALLED}$ = 150.4 kW). The design data on energy consumption for all groups depending on the operating modes are shown in table 1. The data show significant electricity consumption in shore electric power supply mode - 332.63 kW or 37% of the total installed power. The shore supply of the studied yacht is carried out through a shore supply system of type 2.2 (without frequency conversion) installed on board with a nominal power of 400 kVA. The switching of the transformer stages is automatic, performed by a programmable logic controller Simatic S7. The energy balance relative to the actually consumed electric power and energy is shown in figure 5. The shore power supply amounts to 26% of the total energy consumption [4, 5].

Table 1. Total Electricity Consumption

	Regime		Sea going	Shore	Maneuring
№	Consumer	P_{INST} [kW]	P_W [kW]	P_W [kW]	P_W [kW]
1	Group №1	250.2	47.39	15.81	221.8
2	Group №2	412.14	233.33	227.51	231
3	Group №3	82.56	3.39	9.31	17.31
4	Group №4	150.4	88.7	80	83.9
Total		895.3	372.8	332.63	554.01

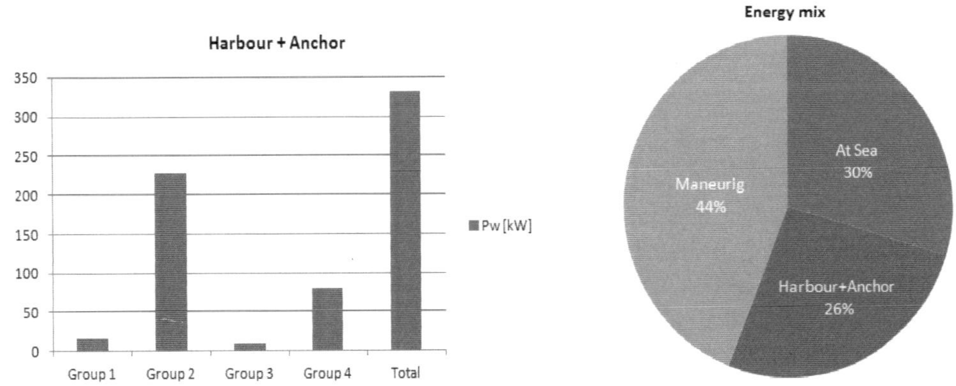

Figure 5. Energy balance and energy mix in shore power supply

4. Conclusion

The report discusses the advantages and disadvantages of using different types of shore power supplies for superyachts. An analysis of an on-board shore-type power supply system for the GT1560 superyacht is presented. The results show a significant share of energy consumption in shore supply mode - about 26% of total energy consumption. The use of onshore supply instead of generators on board the ship leads to an identical reduction in emissions. It should be noted that the onshore supply regime is not specified in the definitions for Energy Efficiency Operational Indicator (EEOI) and Energy Efficiency Design Index (EEDI), as in this case there is no transport work. That is resulting into quantitative value of indexes which become mathematical uncertainty. The report shows two significant issues requiring future research - improving port infrastructure to meet the needs of the growing number of superyachts and the problem of assessing their environmental impact in ports, which is not possible with the standard definition of EEE and EEDI.

5. References

[1] Sandel S and Segerlind J 2016 *Ship propulsion using wind, batteries and diesel electric machinery*, Chamlers University of Technology, Gothenburg, Sweden

[2] Kumar D and Zare F 2019 A comprehensive review of maritime micro grids: system architectures, energy efficiency, power quality and regulations, Digital Object Identifier 10.1109/ACCESS.2019.2917082

[3] Caughpan S and Reynolds K 2016 *Energy efficiency white paper*, Innovative Marine Solutions, US Department of Transportation

[4] Tran T 2020 *The Novelty Numerical Simulation Method for Reducing the Fuel Oil Consumption and the Greenhouse Gas Emission in Shipping Transportation Industry (Cogent Environmental Science* vol 6*)* (Cogent: Taylor & Francis Group) pp 1-24

[5] Sun C, Wang H, Liu C and Zhao Y 2020 *Real Time Energy Efficiency Operational Indicator (EEOI): Simulation Research from the Perspective of Life Cycle Assessment (Journal of Physics: Conference Series* vol 1626) (IOPScience) pp1-8

Acknowledgments

The project discussed herein is supported by the European Regional Development Fund through the Operational Programme "Science and Education for Smart Growth" under contract UNITe № BG05M2OP001-1.001-0008. Project name: "National center of mechatronics and clean technologies".

Statistical models expressing relations between soil moisture, aggregate speed, and tillage depth at plowing and cultivation

P M Veleva[1] and G M Hristova[2]

[1] Trakia University, Department of Agricultural Engineering, Stara Zagora, Bulgaria, petyaveleva26@gmail.com
[2] Trakia University, Department of Agricultural Engineering, Stara Zagora, Bulgaria, galina_jam@yahoo.com

E-mail: petyaveleva26@gmail.com

Abstract. The study is based on a one-year field experiment (2019) in the land of the Chirpan region located in central Bulgaria. The agrotechnical operations of plowing and cultivation, applied in technology for the production of sunflower, are studied. Four models (Linear, Exponential, Logarithmic, and Quadratic) were compared at $p < 0.05$, defining the relation between soil moisture, aggregate speed, and the uniformity of the soil index Tillage depth during plowing and cultivation. It was found that in plowing at a speed of 4 km/h the Quadratic model described the relation between soil moisture and tillage depth with the highest coefficient of determination ($R^2 = 0.682$). Relating to plowing at a speed of 4.5 km/h the most suitable is the Exponential model ($R^2 = 0.729$), i.e. about 68.2% and 72.9% of the variations in tillage depth are due to the influence of the moisture of the soil. The coefficients of determination, calculated when cultivating at speeds of 8 km/h ($R^2 = 0.526$) and 9 km/h ($R^2 = 0.557$), show that the Quadratic model most strongly (52.6% and 55.7%) determines the relation between soil moisture and tillage depth. The developed models could be used to optimize the control systems of agricultural machinery.

1. Introduction

To meet the agro-technical requirements necessary for the production of trench crops, it is not enough to apply only single tillage. Depending on the soil and climatic conditions, it is necessary to carry out several successive treatments differing in time and method of implementation. Long-term research on the application of agrotechnical practices shows the possibility of the approach in choosing a given agro-technical solution [1, 2, 3]. Plowing and cultivation are part of the soil preparation involved in the technology of sunflower production. Depending on the type of soil, during plowing, the machine-tractor unit turns the soil layer to a depth of 25-40 cm. When cultivating, the soil is loosened, mixed, and saturated to a depth of 15 cm. The uniformity of the depth of cultivation is important for restoring the compaction of the soil, improving the penetration of water into the lower layers, changing the distribution of soil aggregates, has a beneficial effect on the growth and development of plants, providing an even sowing bed [4, 5].

Several authors have developed mathematical models, by studying how the depth of tillage influences the growth, development, and productivity of different crops, as well as how affects the physical properties of the soil, its structure, porosity, etc. [1, 4, 6, 7, 8, 9]. Baoyi Ji et al. [10] applied

ANOVA and Duncan's test, finding statistically significant differences between several tillage systems at different depths and tracking the activity of soil enzymes. Panasiewicz et al. [11] evaluated the effect of three tillage systems on the productivity of different crops by using ANOVA, Tukey's, LSD Multiple Range test. The relationship between the parameters was determined with the Pearson correlation coefficient. Ahmad et al. [12] also analysed the effect of deep tillage on soil properties and wheat yield using the General Linear Model procedure. Frink et al. [13], Dallev and Ivanov [4] compile linear and nonlinear models that could be used as a basis for developing systems for optimizing tillage machines.

The information from the available literature on the uniformity of the depth of cultivation in the technological operations of plowing and cultivation depending on the soil moisture is scarce; therefore the expansion of the studies in this direction is justified.

The purpose of this article is to monitor the stability of the machine-tractor unit in terms of depth of tillage and cultivation and to compile statistical models expressing the relations between soil moisture, speed of the aggregate, and tillage depth. These models could be used to optimize the management systems of agricultural machinery.

2. Materials and methods

2.1. Experimental design

The experiments were carried out in the 2019-2020 agricultural year on a field, prepared for sowing sunflowers. The field is located in the Chirpan region in the central part of the Republic of Bulgaria. The area over which the research was conducted is 42 400 m^2 (Figure 1). In the studied area, several types of soils are located unevenly. Humus-carbonate soils predominate for the most part, but there are also traces of resins (leached, typically clayey) and resins (leached, cinnamon, and slightly clayey).

The field on which the experiments were conducted has been divided into two equal sections both with dimensions of 530 m long and 40 m wide (Figure 1). The plowing is done with a five-body reversible plow aggregated to a tractor with 200 hp. at a speed of 4 km/h in the first section, and 4.5 km/h at the second. When cultivating, the machine-tractor unit consisted of a tractor (220 hp.) and a nine-meter cultivator operating at speeds of 8 km/h and 9 km/h accordingly in the first and the second field.

Figure 1. Experimental sections, Chirpan region, Bulgaria (in red squares).

Table 1 presents the period of execution and the depth of processing of the considered technological operations of plowing and cultivation. The depth of processing for the individual operations at the respective speeds of each field is measured with a linear meter, as follows: length - 10 measurements every 50 m; in width - 5 measurements every 8 m. Data on soil moisture has been taken with AM-128 SOIL-soil moisture meter at the respective points of the two sections.

Table 1. Period of execution and depth of processing of the technological operations applied in sunflower production.

Technological operation	Date of performance	Tillage depth (cm)
	Agricultural year (2019-2020)	
Plowing	14.12.2019	28-32
Cultivation	28.01.2020	14-16

2.2. Statistical data analysis

The effect of soil moisture on the depth of tillage for the agrotechnical operations plowing and cultivation at different speeds of the aggregate was analysed using Two-way analysis of variance (ANOVA) and Regression analysis. The aim was to compile predictive models defining the relations between the observed parameters. Level of significance $p < 0.05$ was considered statistically significant. Levine's test of equality of error variances was applied to check the homogeneity of the samples.

Four different regression models - Linear (1), Logarithmic (2), Quadratic (3), and Exponential (4) were compared, given as:

$$a = const.x + b \quad (1)$$

$$a = const. + b\log(x) \quad (2)$$

$$a = const. + b_1x + b_2x^2 \quad (3)$$

$$a = const.e^{bx} \quad (4)$$

where a is the observed parameter tillage depth for agrotechnical operations plowing and cultivation at different speeds, x is the fixed factor soil moisture, and $const.$; b; b_1; b_2 are the model coefficients. The results were calculated with SPSS 26.0.0.1 (IBM© SPSS Statistics, 2019).

3. Results and discussion

The uniformity of the depth of the cultivations (plowing and cultivation) according to the instantaneous soil moisture during the movement of the machine - tractor unit at two speeds is traced and regression models are developed expressing the connection between them.

3.1. Basic statistics and Correlation analysis

Table 2 presents the results of the bivariate correlation method for the observed parameters depth of plowing, speed of the aggregate, and soil moisture at a level of significance $p < 0.05$. The Pearson correlation coefficient (r) reveal that there is a very weak correlation between the factors speed of the aggregate and soil moisture (r = 0.081), i.e. their influence on the studied parameter depth of plowing can be analyzed separately. The relation between the depth of plowing and speed of the aggregate is very weak (r = 0.082). The relationship between the parameters depth of plowing and soil moisture (r = 0.836) is strong and positive, which is a prerequisite for the creation of different regression models and the analysis of the influence of soil moisture on the depth of plowing.

The calculated correlation coefficients of the studied parameters depth of cultivation, speed of the aggregate, and soil moisture at a significance level $p < 0.05$ are presented in Table 3. Similar to the previous results (Table 2), there is a very weak correlation between the factors speed of the aggregate and soil moisture (r = 0.010), i.e. their influence on the studied parameter depth of cultivation can also be examined separately. Regarding the parameters depth of cultivation and speed of the aggregate, a very weak and negative correlation has been registered (r = -0.033). A strong positive correlation is observed between the parameters depth of cultivation and soil moisture (r = 0.727), which confirms the need to create different regression models.

Table 2. Crosstab correlation between examined parameters Depth of plowing, Speed of the aggregate, and Soil moisture.

Tillage depth (a, cm) n=50	**Plowing**	**Aggregate speed (V, km/h)**	**Soil moisture (W, %)**
Plowing	1	0.082	0.836*
Aggregate speed (V, km/h)	-	1	0.081
Soil moisture (W, %)	-	-	1

*Correlation is significant at $p < 0.05$

Table 3. Crosstab correlation between examined parameters Depth of cultivation, Speed of the aggregate, and Soil moisture.

Tillage depth (a, cm) n=50	**Cultivation**	**Aggregate speed (V, km/h)**	**Soil moisture (W, %)**
Cultivation	1	-0.033	0.727*
Aggregate speed (V, km/h)	-	1	0.010
Soil moisture (W, %)	-	-	1

*Correlation is significant at $p < 0.05$

The average values of the observed parameter depth of cultivation of the agro-technical operations plowing and cultivation at the respective speeds are presented in Table 4. As it can be seen from the table, there are no statistically significant differences between the average values of the depth of plowing at a unit speed of 4 km/h (28.17 cm) and 4.5 km/h (28.40 cm). The same results can be seen for the average values of the depth of cultivation at unit speed 8 km/h (15.11 cm) and 9 km/h (15.05 cm). This could be explained by the reported weak correlations between these parameters (Table 2).

Table 4. Basic statistics and ANOVA of the observed parameter Depth of tillage for agrotechnical operations plowing and cultivation.

Tillage depth (a, cm)			
Plowing (n= 50)		Cultivation (n= 50)	
Aggregate speed (km/h)	$\bar{x} \pm SD$	**Aggregate speed (km/h)**	$\bar{x} \pm SD$
4	28.17 ± 1.30^{ns}	**8**	15.11 ± 0.91^{ns}
4.5	28.40 ± 1.52^{ns}	**9**	15.05 ± 0.95^{ns}

^{ns} - Same superscripts within the same column represent insignificant differences at $p < 0.05$; SD – Standard deviation; n – number of the observations

3.2. Regression models

The results of the composed regression models showing the influence of soil moisture on the depth of tillage at plowing and cultivation at different speeds of the aggregate are presented in Tables 5 and 6, Figure 2a-b and Figure 3a-b. Table 5 presents the regression equations showing the effect of soil moisture on plowing depth at a unit speed of 4 km/h. All regression models are statistically significant at $p < 0.05$. The value of the coefficient of determination of the Quadratic model is the highest (R^2 = 0.682), i.e. about 68.2% of the variations in the depth of plowing are explained by the influence of the soil moisture at the speed of the aggregate unit 4 km/h.

Table 5. Model summary and parameter estimation showing the dependence of Depth of plowing (a, cm) from Soil moisture at speed of the aggregate 4 km/h and 4.5 km/h.

Parameter estimates	R^2	Sig.	Equations
		V = 4 km/h	
Linear	0.673	0.001	$a = 0.53x + 16.27$
Logarithmic	0.668	0.003	$a = 7.97 + 11.63\log(x)$
Quadratic	**0.682**	**0.000**	$\boldsymbol{a = 37.33 - 1.39x + 0.04x^2}$
Exponential	0.675	0.000	$a = 18.42e^{0.02x}$
		V = 4.5 km/h	
Linear	0.720	0.002	$a = 0.62x + 14.36$
Logarithmic	0.717	0.001	$a = -14.13 + 13.62\log(x)$
Quadratic	0.721	0.000	$a = 19.96 + 0.11x + 0.01x^2$
Exponential	**0.729**	**0.000**	$\boldsymbol{a = 17.23e^{0.02x}}$

* Level of significance $p < 0.05$

The calculated regression equations showing the influence of the soil moisture on the depth of plowing at a unit speed of 4.5 km/h were statistically significant at $p < 0.05$ (Table 5). In this case, the value of the coefficient of determination of the Exponential model is the highest ($R^2 = 0.729$), i.e. the parameter soil moisture at the speed of the aggregate 4.5 km/h affects the depth of plowing about 72.9%. Figures 2a and 2b show the estimation of the curves of the developed regression models expressing the relation between soil moisture and the depth of plowing at the respective speeds. The Quadratic model best describes the studied dependences at the speed of the unit 4 km/h (Figure 2a), and the Exponential model respectively at speed of 4.5 km/h (Figure 2b).

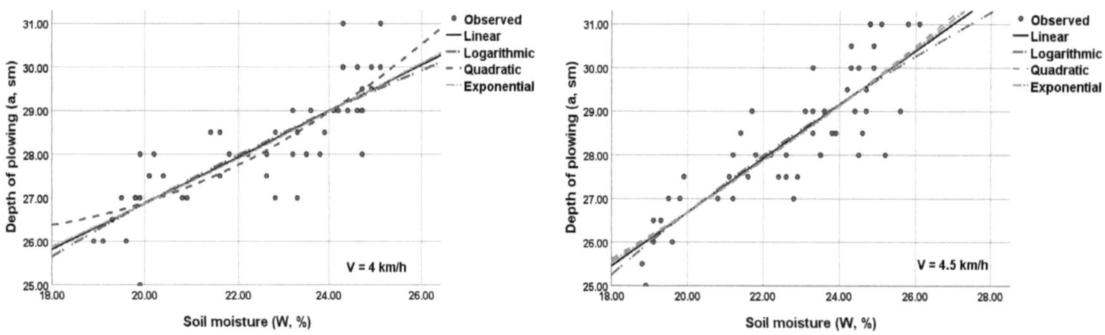

a) b)

Figure 2. Curve estimation of regression models showing the influence of the Soil moisture on the Depth of plowing.

The regression equations showing the influence of soil moisture on the depth of cultivation at speeds of the aggregate 8 km/h and 9 km/h are presented in Table 6. All models are statistically significant at $p < 0.05$. The values of the coefficient of determination for both examined speeds are the highest for the Quadratic model ($R^2 = 0.526$; 0.557), i.e. about 52.6% of the variations in the variable depth of cultivation are explained by the influence of soil moisture at speed of the aggregate 8 km/h and 55.7% at speed of 9 km/h.

Table 6. Model summary and parameter estimation showing the dependence of Depth of cultivation (a, cm) from Soil moisture at speed of the aggregate 8 km/h and 9 km/h.

Parameter estimates	R^2	Sig.	Equations
		V = 8 km/h	
Linear	0.523	0.000	$a = 0.57x + 6.04$
Logarithmic	0.519	0.001	$a = -10.05 + 9.11\log(x)$
Quadratic	**0.526**	**0.000**	$\boldsymbol{a = 16.14 - 0.69x + 0.04x^2}$
Exponential	0.515	0.001	$a = 8.32e^{0.04x}$
		V = 9 km/h	
Linear	0.535	0.004	$a = 0.58x + 5.82$
Logarithmic	0.526	0.001	$a = -10.29 + 9.17\log(x)$
Quadratic	**0.557**	**0.000**	$\boldsymbol{a = 29.85 - 2.44x + 0.09x^2}$
Exponential	0.523	0.002	$a = 8.18e^{0.04x}$

* Level of significance $p < 0.05$

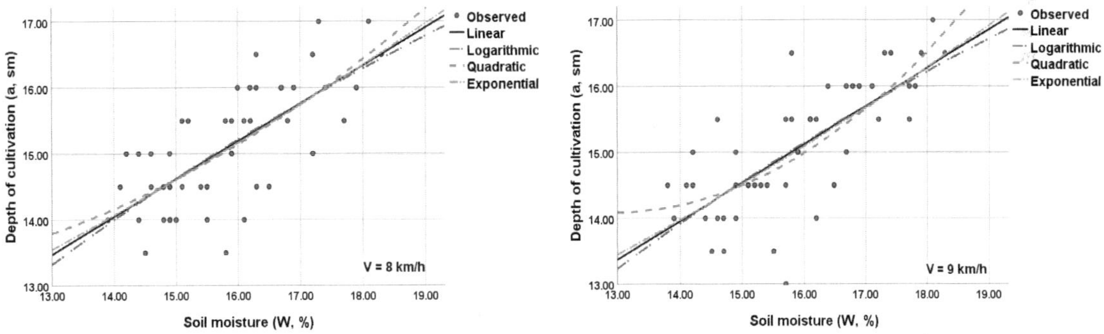

a) b)

Figure 3. Curve estimation of regression models showing the influence of the Soil moisture on the Depth of cultivation.

The estimation of the curves of the models determining the relation between soil moisture and depth of cultivation at speeds of 8 km/h and 9 km/h is presented in Figures 3a and 3b. The Quadratic model best describes the studied dependences for both observed speeds.

4. Conclusions

Strong, positive correlations were reported between the parameters Depth of plowing, Depth of cultivation, and Soil moisture (r = 0.836; 0.727), which justifies the creation of regression models.

The average value of depth of plowing at a unit speed of 4 km/h is 28.17 cm, and at 4.5 km/h is 28.40 cm. The average value of depth of cultivation at a unit speed of 8 km/h is 15.11 cm, and at 9 km/h is 15.05 cm. It was found that the different speeds did not have a statistically significant effect on the depth of tillage during plowing and cultivation.

The Quadratic model best describes the influence of soil moisture on the depth of plowing at a speed of the aggregate 4 km/h (R^2 = 0.682), and the Exponential model respectively at a speed of 4.5 km/h (R^2 = 0.729). The influence of soil moisture on depth of cultivation for both observed speeds (8 km/h and 9 km/h) is best defined with the Quadratic model (R^2 = 0.526; 0.557).

The developed models could be used in the optimization of the control systems of the machine-tractor units and to serve the farmers in the choice of some parameters for carrying out certain technological tillage.

5. References

[1] Dimitrov I, Nenov M, Traikov N, Toncheva R and Gerasimova I 2016 Agrotechnical impacts on the physical characteristic of the soil and ways to remove the negative changes. I. Under non-irrigation conditions *Soil Science Agrochemistry and Ecology* **50** 3-4 169-178

[2] Nenov M, Dimitrov I, Lozanova V, Gerasimova I and Traikov N 2020 Optimization of agrotechnical units in technologies for grain production in the Sofia area *Bulgarian Journal of Soil Science Agrochemisty and Ecology* **54** 3 28-40

[3] Ulrich S, Hofmann B, Tischer S and Christen O 2006 Influence of tillage on soil quality in a long – term trial in Germany. *Soil management for sustainability* ed A Horn, H Fleig, S Peth and X Peng (Reiskirchen: Catana Verlag) pp 110-116

[4] Dallev M and Ivanov I 2015 Study of body for surface tillage in heavy soils with low humidity *Scientific Papers. Series A. Agronomy* **LVIII** 45-48

[5] Dallev M, Ivanov I and Popova R 2017 Investigation of control erosion by surface treatment of the soil with tillage active body *52nd Croatian & 12th International Symposium on Agriculture* 637-640

[6] Dunker R, Hooks C, Vance S and Darmody R 1995 Deep tillage effects on compacted surface-mined land *Soil Science Society of America Journal* **59** 192-199

[7] Dimitrov I 2014 Criteria of the choice of a level of the soil tillage in dependence of some physical parameters *Soil science agrochemistry and ecology* **XLVIII** 1 21-25

[8] Tian P, Lian H, Wang Z, Jiang Y, Li C, Sui P and Qi H 2020 Effects of deep and shallow tillage with straw incorporation on soil organic carbon, total nitrogen and enzyme activities in Northeast China, *Sustainability* **12** 8679

[9] Laddha K C and Totawat K L 1997 Effects of deep tillage under rainfed agriculture on production of sorghum (*Sorghum biocolor L. Moench*) intercropped with green gram (*Vigna radiata* L. Wilczek) in western India *Soil and Tillage Research* **43** 241-250

[10] Ji B, Hu H, Zhao Y, Mu X, Liu K and Li C 2014 Effects of Deep Tillage and Straw Returning on Soil Microorganism and Enzyme Activities *The Scientific World Journal* 451493

[11] Panasiewicz K, Faligowska A, Szymanska G, Szukała J, Ratajczak K and Sulewska H 2020 The effect of various tillage systems on productivity of Narrow-Leaved Lupin-Winter Wheat-Winter Triticale-Winter Barley rotation *Agronomy* **10** 304

[12] Ahmad M, Abdullah H, Iqbal M, Umair M and Ghani M U 2010 Effect of deep tillage on soil properties and crop (wheat) yield *Soil and Environment* **29** (2) 177-180

[13] Frink E, Flippo D and Sharda A 2016 Primitive geometry tillage modeling, *Open Journal of Soil Science* **6** 34-43

Evaluation of the behaviour of an induction motor for fan system under various external factors

S Rachev[1], **L Dimitrov**[1]

[1] Department of Electrical Engineering, Technical University of Gabrovo, 4 Hadgi Dimitar str., Gabrovo 5300, Bulgaria

E-mail: sratchev@mail.com

Abstract. Paper concerns a mathematical model developed for working process observation of a fan system squirrel-cage induction motor electric drive. The electric motor in question is designed for permanent, uninterruptible operation. The system of differential equations is converted and processed with the help of appropriate software. The impact of external factors is assessed – supply voltage deflection and variation of rates of mechanical quantities involved in the motion equation. Electrical power losses in steady-state regime and energy losses at start-up are calculated. The results obtained are practical oriented when considering methods to start and control the rotational frequency to imrove energy efficiency.

1. Introduction

According to global studies, the largest consumers of electricity produced in the world are electric motors. The largest share is occupied by induction motors, also called asynchronous. According to expert's opinion, they represent 80-90% of all electric motors used. It has been found that about 55% of induction motors are used to drive mechanisms such as fans and pumps.

Fan systems are widely used. The term "fans" includes generators for compressed gases, most often for compressed air. Their division into groups depends on the pressure of the exhaust gases and their purpose. The fans create a difference in outlet and inlet pressure from 0.001 to 0.01 MPa. Their purpose is for ventilation of industrial buildings and facilities, for heating systems (air conditioning systems), for cooling motors and rectifiers, for suction of flue gases during central and local ventilation in shafts, for pneumatic transport, for agriculture, etc.

Fans are an inseparable part of different technological processes in the metallurgy, chemical and energy industries. They also serve to creating of technologically clean environment – for production of elements in radio-electronics and computer machinery, in laboratories for tissue cultivation. Practically, there is no branch of every national economy, where fans don't find application.

By design, fans are divided into centrifugal and axial. Fans are slow-moving and heavy mechanisms. The fans operate with a low degree of compression, and centrifugal fans are used to move lower flow rates at higher pressures than axial fans.

The fans are working hydraulic turbo-mechanisms which convert the mechanical energy into hydraulic – kinetic and potential. The main energy changing component is the impeller, which can be radial, diagonal and axial. Generally the fans with radial and diagonal impellers are called centrifugal fans [1].

Fan motor power ranges from a few tens of watts to several thousand kilowatts. They are usually started without load and the conditions for starting the motors are light. The operating mode is continuous and the the number of releases (starts) - small.

The productivity, reliability of operation, simplicity of service and automation potential of the mechanisms is in direct dependence on the advanced technical characteristics of electric drives. For this reason, a detailed examination of the electric drive of mechanisms and machines is of essential practical importance. This also applies in full force to the fan drives.

2. Technical considerations

The following should be considered when selecting a fan [2]:

- Type of site – industrial, commercial, residential
- Fluid characteristics – clean air, air with dust or grease, flue gases, operating temperature, etc.
- Type of ventilation system – injection or suction;
- Required flow and pressure;
- Type of power supply – single-phase or three-phase;
- Noise level conditions.

The fans general parameters are:

- Flow rate Q - the volume of air passed through the fan per unit time; measured in $[m^3/s]$ or $[m^3/h]$;
- Full pressure P_t $[Pa]$ – shows the increase in the specific energy of the air passed through the fan;
- Dynamic pressure P_d $[Pa]$ – characterizes the air kinetic energy after the fan;
- Static pressure P_{st} $[Pa]$ – represents the difference between P_t and P_d ;
- Power – represents the electric motor full power, providing the fan parameters; measured in $[W]$ or $[kW]$.

When ordering fan to the manufacturer, the following characteristics should be stated: type of fan; capacity [m^3/h]; installed power; rotational frequency per minute [min^{-1}]; characteristics of the fluid – temperature, explosiveness, dustiness, adhesiveness of the dust and other specific requirements; fan equipment – frame, vibrated frame, flexible joints and other [1].

Feature of fans such as the so-called turbo-mechanisms is the dependence of the static resistance moment on the angular velocity squared.

In order for a fan to start normally, it is necessary for the starting torque of the drive motor to be greater than the resting moment of the turbo-mechanism, which is determined by the aerodynamic (or hydraulic) drag, but not to exceed 115-120% of the latter one.

Fans rank second among turbo-mechanisms after industrial pumps. Fans operating in production facilities consume significant total electricity.

For powers up to 250 kW, the fans are driven mainly by squirrel cage induction motors, and above this value - by synchronous motors.

Of particular importance for a number of fans is the application of adjustable drive. At present, the trends of the transition to adjustable fan drive have become more than obvious. The accumulated experience from the operation points out that the efficiency of the fans with adjustable electric drive is higher than with non-adjustable motors, on average by 12%.

In direct connection with energy efficiency is the observance of the limits of the so-called indicators of electricity quality. An assessment of the impact of electricity consumers on the overall energy efficiency through an implemented methodology is proposed in [3]. In order to improve energy efficiency, it is possible to define specific measures. In addition, modern control methods are sources of harmonic distortions that "pollute" the mains and negatively affects the normal operation of other consumers [4].

There is no universal criterion for optimality. Reliability of operation in general and improvement of energy performance are important goals for the electric motors used. In fact, for example, in conventional design methods, minimizing motor mass results in lower energy performance and reduced reliability.

Below are recommended values for different cases – the change of air in a certain space with what frequency for an hour to happen [2]:

For industrial purposes: boiler rooms: 20÷25 times; fabric dyeing rooms: 10÷15 times; powder coating rooms: 10÷15 times; machine rooms: 20÷30 times; factory halls: 3÷6 times; foundries: 30÷40 times; painting rooms: 30÷60 times; welding rooms: 15÷30 times.

For commercial purposes and others: bakeries: 20÷30 times; bank offices: 3÷4 times; coffee bars: 10÷15 times; dining rooms: 5÷10 times; cinemas and theaters: 5÷8 times; conference rooms: 8÷12 times; couloirs: 3÷5 times; garages: 6÷8 times; sports halls: 6÷12 times; hair salon: 10÷15 times; toilets: 8÷15 times; offices: 4÷8 times; libraries: 3÷5 times; recording studios: 10÷12 times; restaurants: 6÷10 times; school rooms: 2÷4 times.

These values are optional and may vary from case to case. The flow rate required is related to the quantity of air needed for one person (these are minimum values): at normal activity: 20÷25 m^3/h; during normal activity with smoking allowed: 30÷35 m^3/h; for light physical work: 45 m^3/h.

The flow rate can be determined by means of the velocity of pull-out:

- For home kitchens: 0.15 up to 0.20 *m/s*; for industrial kitchens: 0.20 up to 0.25 *m/s*.
- When degreasing: 0.25 up to 0.50 *m/s*; When galvanizing: 0.50 up to 1 *m/s*; For aerosol chambers: 0.70 up to 1 *m/s*.

The flow rate can also be calculated from the transportation speed: dust: 9 *m/s*; flour: 13 *m/s*; sawdust: 15 *m/s*; metal powder: 15 *m/s*; shavings: 18 m/s; lead powder: 20÷25 *m/s*. To calculate the flow, the speed must be multiplied by the cross section of the pipeline.

3. Mathematical model

Centrifugal double-inlet fan VCP-8 intended to complete fan section from air-conditioning chambers has been model studied. It transports air and gases, free from mechanical particles, with a maximum temperature up to 223.16 °K (50 °C), by providing a flow rate Q=11000-27000 m^3/h and full pressure Pt=300-4000 Pa. It is completed with an electric motor and starting equipment with grade of protection IP44, belts and slide rails for the electric motor.

The mathematical model applied is similar to these ones in [5, 6, 7].

The model voltages of the windings are as follows [5, 6]:

$$\begin{aligned} u_{sx} &= r_1 i_{sx} + \frac{d\Psi_{sx}}{dt} - \Omega_s \Psi_{sy}; \\ u_{sy} &= r_1 i_{sy} + \frac{d\Psi_{sy}}{dt} + \Omega_s \Psi_{sx}; \\ u_{rx} &= r_2' i_{rx} + \frac{d\Psi_{rx}}{dt} - s\Omega_s \Psi_{ry}; \\ u_{ry} &= r_2' i_{ry} + \frac{d\Psi_{ry}}{dt} + s\Omega_s \Psi_{rx} \end{aligned}, \qquad (1)$$

where

$\Omega_s = \frac{2\pi f_N}{p} = \frac{\omega}{p} = \frac{\omega_b}{p}$ – field synchronous angular speed; $\omega = 2\pi f_N = \omega_b$ – circular frequency, accepted as a basic one.

p – number of pole pairs; u_{sx}, u_{sy} – model stator voltages; u_{rx}, u_{ry} – model stator voltages;

r_1, r_2' – stator and reduced rotor ohmic resistances; $i_{sx}, i_{sy}, i_{rx}, i_{ry}$ – model stator and rotor currents

$\Psi_{sx}, \Psi_{sy}, \Psi_{rx}, \Psi_{ry}$ – magnetic fluxes of the stator and rotor; s – the electric motor slip.

We use the relative units system according the generally accepted basic values (for example $\tau = \omega_b t$ - time in relative units). The parameters of the electric motor substitution circuit are determined for a value of the slip $s = 1$ by the calculation method [8]. Some technical data and parameters of the electric motor and fan are given in Appendix.

$$\frac{di_{sx}^*}{d\tau} = \frac{L_r^* . L_m^*}{L_e^*} \left[\begin{array}{l} \frac{u_{sx}^*}{L_m^*} - \frac{u_{rx}^*}{L_r^*} - \frac{r_1^*}{L_m^*} . i_{sx}^* + \frac{r_2^*}{L_r^*} . i_{rx}^* + ... \\ .... + i_{sy}^* . \left(\frac{L_s^* . L_r^* - s . L_m^{*\,2}}{L_r^* . L_m^*} \right) + (1-s) i_{ry}^* \end{array} \right] \quad \frac{di_{sy}^*}{d\tau} = \frac{L_r^* . L_m^*}{L_e^*} \left[\begin{array}{l} \frac{u_{sy}^*}{L_m^*} - \frac{u_{ry}^*}{L_r^*} - \frac{r_1^*}{L_m^*} . i_{sy}^* + \frac{r_2^*}{L_r^*} . i_{ry}^* + ... \\ .... + i_{sx}^* . \left(\frac{s . L_m^{*\,2} - L_s^* . L_r^*}{L_r^* . L_m^*} \right) - (1-s) i_{rx}^* \end{array} \right]$$

$$\frac{di_{rx}^*}{d\tau} = \frac{L_s^* . L_m^*}{L_e^*} \left[\begin{array}{l} -\frac{u_{sx}^*}{L_s^*} + \frac{u_{rx}^*}{L_m^*} + \frac{r_1^*}{L_s^*} . i_{sx}^* - \frac{r_2^*}{L_m^*} . i_{rx}^* + ... \\ ... + i_{ry}^* . \left(\frac{s . L_s^* . L_r^* - L_m^{*\,2}}{L_s^* . L_m^*} \right) - (1-s) i_{sy}^* \end{array} \right] \quad \frac{di_{ry}^*}{d\tau} = \frac{L_s^* . L_m^*}{L_e^*} \left[\begin{array}{l} -\frac{u_{sy}^*}{L_s^*} + \frac{u_{ry}^*}{L_m^*} + \frac{r_1^*}{L_s^*} . i_{sy}^* - \frac{r_2^*}{L_m^*} . i_{ry}^* + ... \\ .... + i_{rx}^* . \left(\frac{L_m^{*\,2} - s . L_s^* . L_r^*}{L_s^* . L_m^*} \right) + (1-s) i_{sx}^* \end{array} \right] \quad (2)$$

where $L_s^*, L_r^*, L_m^*, L_e^*$ – stator, rotor, mutual and equivalent inductance in relative units, respectively.

The stator voltages u_{sx}^* and u_{sy}^* are as follows [5]:

$$\begin{array}{l} u_{sx}^* = \sqrt{\frac{2}{3}} k_V \left[u_A^* \cos\tau + u_B^* \cos(\tau - \frac{2\pi}{3}) + u_C^* \cos(\tau + \frac{2\pi}{3}) \right]; \\ u_{sy}^* = -\sqrt{\frac{2}{3}} k_V \left[u_A^* \sin\tau + u_B^* \sin(\tau - \frac{2\pi}{3}) + u_C^* \sin(\tau + \frac{2\pi}{3}) \right] \end{array} \quad (3)$$

where u_A^*, u_B^* and u_C^* are the phase voltages in relative units;

$k_V = V / V_{NOM}$ gives the account of supply voltage value, so called *multiplicity of supply voltage* [6];

V – supply voltage value in circulation; V_{NOM} – supply voltage rated value.

Because of squirrel-cage rotor relevant voltages are $u_{rx}^* = 0$ and $u_{ry}^* = 0$. The phase stator voltages in relative units are obtained to be:

$$u_A^* = k_V \cos(\tau + \varphi_0); u_B^* = k_V \cos(\tau + \varphi_0 - \frac{2\pi}{3}); u_C^* = k_V \cos(\tau + \varphi_0 + \frac{2\pi}{3}), \quad (4)$$

where φ_0 – initial phase of supply voltage.

It is proved that after substitution of (4) into (3) and transformation the result is:

$$u^*_{sx} = \sqrt{\frac{2}{3}} k_V \frac{3}{2} \cos\varphi_0 = \sqrt{\frac{3}{2}} k_V \cos\varphi_0; \; u^*_{sy} = -\sqrt{\frac{2}{3}} k_V \left(-\frac{3}{2}\right) \sin\varphi_0 = \sqrt{\frac{3}{2}} k_V \sin\varphi_0 \; . \qquad (5)$$

The phase stator currents are as follows:

$$\begin{aligned} i^*_A &= \sqrt{\frac{2}{3}} \left[i^*_{sx} \cos\tau - i^*_{sy} \sin\tau \right] \\ i^*_B &= \sqrt{\frac{2}{3}} \left[i^*_{sx} \cos(\tau - \frac{2\pi}{3}) - i^*_{sy} \sin(\tau - \frac{2\pi}{3}) \right] \\ i^*_C &= \sqrt{\frac{2}{3}} \left[i^*_{sx} \cos(\tau + \frac{2\pi}{3}) - i^*_{sy} \sin(\tau + \frac{2\pi}{3}) \right] \end{aligned} \qquad (6)$$

The induction motor electromagnetic torque is as follows (M_N - base moment):

$$M^* = \frac{M}{M_N} = \frac{pL_b i_b^2 L^*_m}{M_N} \left(i^*_{sy} i^*_{rx} - i^*_{sx} i^*_{ry} \right) = \frac{pu_b i_b L^*_m}{M_N \omega_b} \left(i^*_{sy} i^*_{rx} - i^*_{sx} i^*_{ry} \right). \qquad (7)$$

After conversion, the motion equation is as follows [9]:

$$\frac{ds}{dt} = -\frac{p}{J_\Sigma \omega} (M - M_L), \qquad (8)$$

where M_L – mechanism resistance moment; J_Σ – total inertia moment of the drive system.

The fan blower is a mechanism which characteristics is presented by following equation [9]:

$$M^*_L = M^*_{L_{INIT}} + \left(M^*_{LN} - M^*_{L_{INIT}} \right) (1 - s)^2 \qquad (9)$$

where $M^*_{L_{INIT}}$ – initial resistance moment; M^*_{LN} – rated resistance moment.

After transformation equation (8) is obtained to be:

$$\frac{ds}{d\tau} = -\frac{pM_b}{J_\Sigma \omega_b^2} \left[\frac{pu_b i_b L^*_m}{M_N \omega_b} \left(i^*_{sy} i^*_{rx} - i^*_{sx} i^*_{ry} \right) - M^*_{L_{INIT}} - \left(M^*_{LN} - M^*_{L_{INIT}} \right) (1 - s)^2 \right] \qquad (10)$$

The dynamic behaviour of a fan driven by an induction motor has been studied by a system of five differential equations formed by equations (2) and (10) [7, 10]. The four of these equations are for model stator currents and the fifth one is related to torques and it is the motion equation [7, 11]. As mentioned above, the moment of resistance is changed by so-called 'square-law' [12]. Let us emphasize that the basic equations of any induction machine are strongly nonlinear.

The electric drive total inertia moment J_Σ is presented through factor of inertia FI and rotor moment of inertia J_r [10]:

$$J_\Sigma = FI \times J_r \qquad (11)$$

The variable electrical losses in the stator and rotor, the magnitude of which depends on the values of the flowing currents, are calculated as follows [5, 6, 7]:

$$\Delta P_V = \Delta P_1 + \Delta P_2 \approx 3I_1^2 r_1 + 3I_2^{'2} r_2^{'}, \qquad (12)$$

where

I_1 – stator current; $I_2^{'}$ – rotor current reduced to the stator;

$\Delta P_1 = 3I_1^2 r_1$ – electrical power losses associated with stator windings heating when current flows thereon.

$\Delta P_2 = 3I_2^{'2} r_2^{'}$ – again such losses, but in rotor windings.

If the induction motor operates in the small slip zone to the nominal one, the equality applies $I_1^2 \approx I_2^{'2} + I_0^2$, where I_0 – magnetizing current.

The following is therefore in force:

$$\Delta P_V = 3I_2^{'2}\left(r_1 + r_2^{'}\right). \tag{13}$$

Since

$$I_2^{'} = \sqrt{\frac{M\omega_0 s}{3r_2^{'}}} \tag{14}$$

Th e electrical losses are ultimately represented in such a way:

$$\Delta P_v = \Delta P_1 + \Delta P_2 = M\omega_0 s(1 + \frac{r_1}{r_2^{'}}) \tag{15}$$

where ω_0 – synchronous angular speed.

The differential equations system combining equations (2) and (10) was subsequently treated by the functional calculus Rkadapt with an adaptive approach step size embedded in the Mathcad® software of the PTC® (Parametric Technology Corporation).

4. Results and findings

Through the mathematical model implemented the processes during fan operation have been studied, as the supply voltage and the factor of inertia vary. The combined influence of the two quantities was studied, not separately.

Some of the results are presented in tables and graphs.

Table 1 presents the influence of supply voltage and factor of inertia *FI* on the emerging impact torques and currents. The starting time is calculated, with the criterion that two immediately following values of angular velocity differ by no more than 1%.

This means that electromechanical transient processes have been completed, and electromagnetic transient processes are known to end long before.

Table 2 presents overall electrical power and energy losses calculated. The speed-time and torque-time characteristics for $V = 0.9V_N$ and $FI = 5.5$ are drawn in Figure 1 and Figure 2, separately. The other figures represent: Figure 3 and Figure 4 – dynamic mechanical characteristics in case of Factor of Inertia $FI = 1.5$ and $FI = 5.5$ for $V = 0.9V_N$. The other figures represent: Figure 5 – starting time vs. factor of inertia dependence; Figure 6 – electrical power losses vs. multiplicity of supply voltage; Figure 7 – energy losses vs. factor of inertia dependence; Figure 8 – electrical power losses vs. multiplicity of supply voltage.

Currently, electric motors must meet the recommendations of IEC 60034-30-1: 2014 in terms of efficiency levels (IE codes) or to meet the NEMA MG 1 standards (meet or exceed the recommendations for energy efficiency level Premium Efficiency). From 01.01.2017 in the European Union all electric motors with power from 0.75 *kW* to 375 *kW* must have an efficiency corresponding to IE3 code, except those designed to operate with an electronic speed controller (VSD), which may be IE2 efficient.

The considered centrifugal double-inlet fan VCP-8 is intended to complete fan section from air-conditioning chambers, which is very different from the application for flue gas extraction, for example. Because of higher number of starts there is a possibility to apply frequency-controlled motor with a quality inverter - the so-called variable speed drive (VSD).

For the European Union, the energy efficiency class of such electric motors must be at least IE2 (which is a cheaper technical solution at competitive prices compared to an IE3 motor). With such a drive, the speed, torque and power output can be adjusted. This improves drive efficiency and saves energy.

The speed control of electric motors with VSD is performed in 2 ranges:

- up to rated speed - a constant torque of the motor shaft is maintained, the voltage frequency varying about in the range from 5 *Hz* to 50 *Hz*. The mechanical power of the motor shaft varies from minimum to rated value..
- from rated to maximum speed - constant power is maintained when changing the speed of rotation.

The ratio of maximum torque to rated torque is kept constant.

Table 1. The supply voltage and factor of inertia effects.

K_V	FI	T_{imp}, Nm	i_{imp} *	t_{st}, s
	1.5	5.006	8.068	0.507
	2.5	5.186	8.149	0.423
1.00	3.5	5.257	8.244	0.482
	4.5	5.297	8.832	0.523
	5.5	5.366	8.865	0.570
	1.5	4.563	8.597	0.582
	2.5	4.704	8.637	0.585
0.95	3.5	4.761	8.761	0.591
	4.5	4.791	8.703	0.607
	5.5	4.328	8.983	0.621
	1.5	4.133	7.750	0.384
	2.5	4.242	8.125	0.407
0.90	3.5	4.286	8.185	0.428
	4.5	4.374	8.065	0.506
	5.5	4.489	8.218	0.523
	1.5	5.460	9.502	0.421
	2.5	5.685	9.642	0.438
1.05	3.5	5.774	9.020	0.457
	4.5	5.821	9.897	0.483
	5.5	5.850	9.951	0.521
	1.5	5.293	9.886	0.553
	2.5	6.202	9.893	0.498
1.10	3.5	6.312	9.933	0.492
	4.5	6.370	10.195	0.418
	5.5	6.406	10.289	0.423
	1.5	6.392	10.207	0.706
	2.5	6.775	10.361	0.382
1.15	3.5	6.869	10.634	0.375
	4.5	6.940	10.537	0.362
	5.5	6.984	10.599	0.438
	1.5	6.865	10.499	0.752
	2.5	7.283	10.857	0.423
1.20	3.5	7.446	11.100	0.384
	4.5	7.532	10.770	0.371
	5.5	7.584	10.744	0.408

Table 2. Overall losses of electrical power and energy

K_V	FI	**Electrical power losses – steady state regime** $\Delta P_1 + \Delta P_2, W$	**Energy losses – starting regime** W_{START}, kWh
	1.5		0.774
	2.5		1.291
1.00	3.5	435	1.808
	4.5		2.323
	5.5		3.191
	1.5		0.894
	2.5		1.495
0.95	3.5	482	2.113
	4.5		2.788
	5.5		3.484
	1.5		0.591
	2.5		1.042
0.90	3.5	538	1.533
	4.5		2.329
	5.5		2.940
	1.5		0.638
	2.5		1.114
1.05	3.5	394	1.626
	4.5		2.209
	5.5		2.911
	1.5		0.834
	2.5		1.265
1.10	3.5	363	1.748
	4.5		1.909
	5.5		2.248
	1.5		1.058
	2.5		0.966
1.15	3.5	328	1.329
	4.5		1.650
	5.5		2.440
	1.5		1.420
	2.5		1.071
1.20	3.5	301	1.359
	4.5		1.689
	5.5		2.270

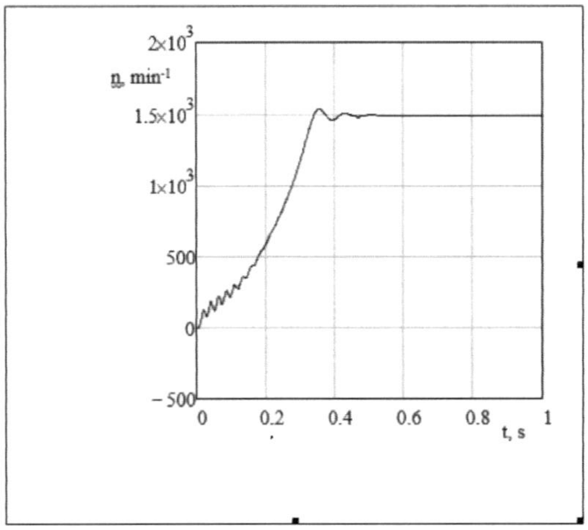

Figure 1. Speed-time characteristic for $U = 0.9U_N$ and $FI = 5.5$.

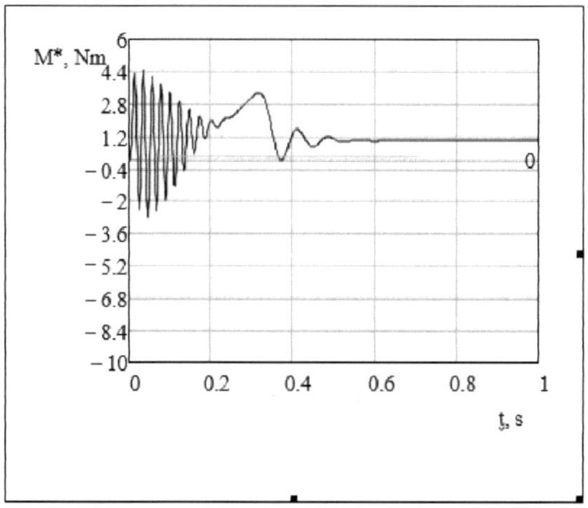

Figure 2. Torque-time characteristic for $U = 0.9U_N$ and $FI = 5.5$.

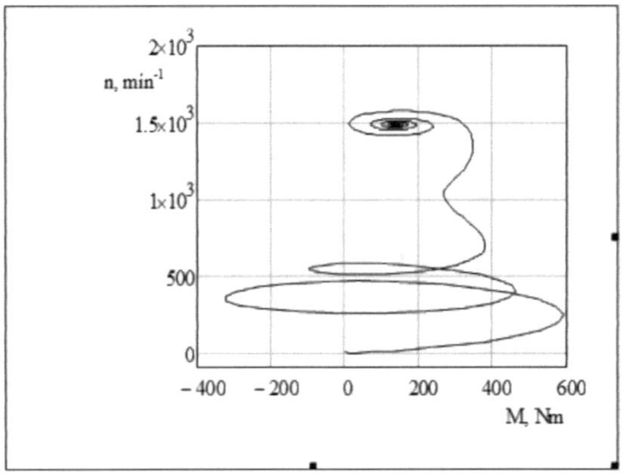

Figure 3. Dynamic mechanical characteristic in case of Factor of Inertia $FI = 1.5$ for $V = 0.9V_N$.

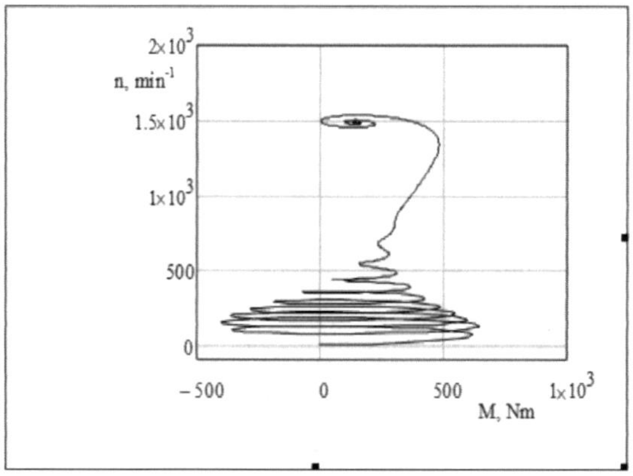

Figure 4. Dynamic mechanical characteristic in case of Factor of Inertia $FI = 5.5$ for $V = 0.9V_N$.

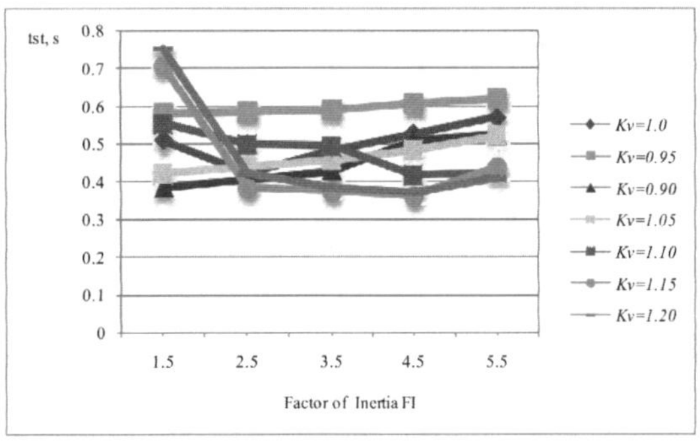

Figure 5. Starting time vs. factor of inertia dependence

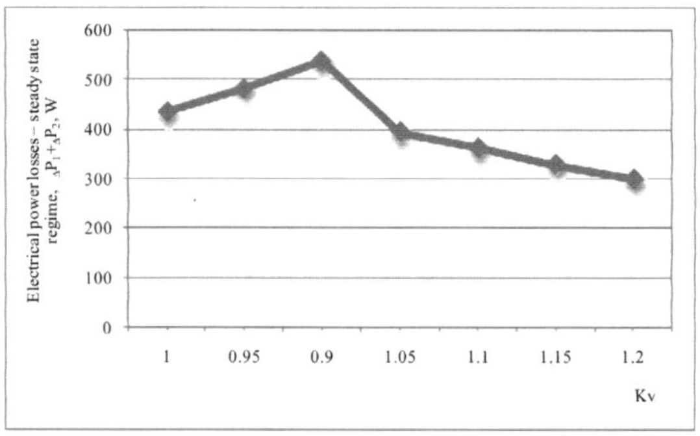

Figure 6. Electrical power losses vs. multiplicity of supply voltage

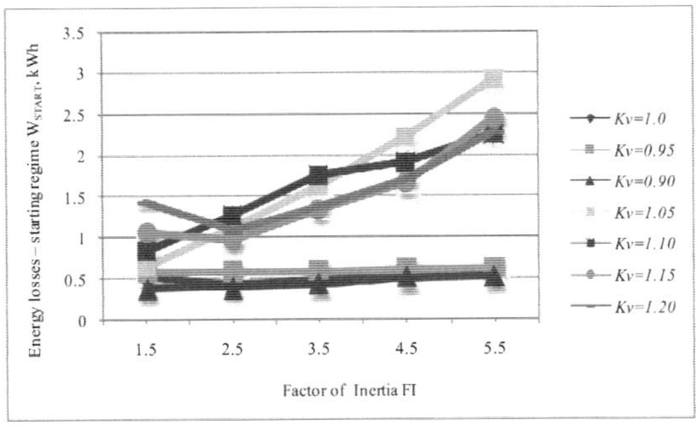

Figure 7. Energy losses vs. factor of inertia dependence

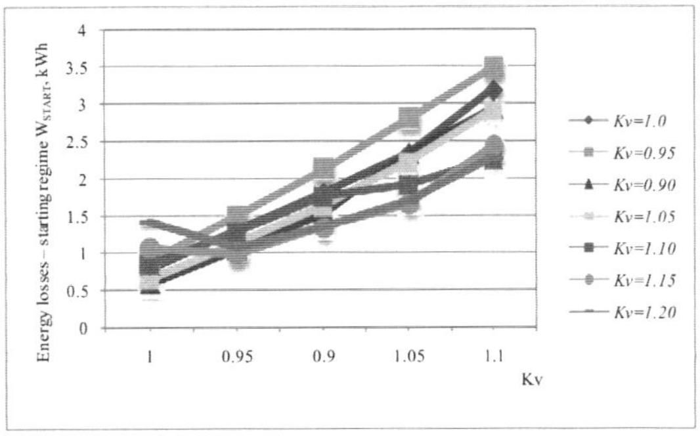

Figure 8. Energy losses vs. multiplicity of supply voltage.

5. Conclusion

With the help of the proposed mathematical model it is possible to analytically study the operation of a fan and in this case to assess the influence of operating voltage and inertial masses on the ongoing electromagnetic and electromechanical processes. Values for the resulting impact torques, currents, angular velocities and accelerations are also obtained.

The stator electrical losses and flowing current are increased in a heavy workload and decreased at low motor load. The share of total electrical losses in relation to the rated power is up to 2.4%.

In addition to theoretical study of electric drives in operation, the developed model can be applied for expedient choice of drive motor and its control, ensuring the behavior of the entire electromechanical system under user-specified requirements and initial conditions.

The results obtained are practical oriented when considering methods to start and control the rotational frequency of induction motors driving turbo-mechanisms.

Appendix

Technical data of the electric motor AT2 180 L4 [14]:

$P_N = 22kW$; $V_N = 400V$; $f_N = 50Hz$; $p = 2$; $n_N = 1450\,min^{-1}$;

$Iph_N = 72.3A$; $J = 0.07646kgm^2$; $I_{ST} / I_N = 7.0$; $M_{ST} / M_N = 2.2$; $M_{MAX} / M_N = 2.9$;

$r_1 = 0.4843\Omega$; $x_1 = 1.154\Omega$; $r_2' = 0.619\Omega$; $x_2' = 1.195\Omega$; $x_m = 51.719\Omega$.

Technical data of fan blower VCP-8, made by Spartak JSC, Burgas, BULGARIA [1]:

$n_N = 1320\,min^{-1}$; $P_N = 22kW$; Sound level pressure 93 dB(A).

6. References

[1] https://www.spartak.bg

[2] http://www.heriks.com

[3] Ivanova G 2020 Specifics in determining the operational energy efficiency index eeoi according to the requirements of the international maritime organization IMO for the period 2020-2025, *12th Electrical Engineering Faculty Conference BulEF 2020*, Varna, Bulgaria

[4] Stoyanov I, Iliev T, Evstatiev B and Mihaylov G 2019 Harmonic distortion by single-phase photovoltaic inverter *11th International Symposium on Advanced Topics in Electrical Engineering (ATEE)* pp 1-4, doi: 10.1109/ATEE.2019.8725009

[5] Hughes A 2005 Electric Motors Drives Fundamentals, Types and Applications Third edition (Newnes 2005) p 384

[6] Rachev S 2004 Dynamic study of fan blower induction motor electric drive, *4th International Conference RADMI 2004*, Zlatibor Serbia and Montenegro 2004, pp 268÷273

[7] Rachev S, Dimitrov L, Ivanov I, Karakoulidis K 2017 Study the effects of no nominal conditions on the performance of high efficiency induction motor, *8th International Conference on Energy and Environment CIEM 2017*, University Politehnica of Bucharest, Romania IEEE *Xplore*

[8] Rachev S, Dimitrov L, Karakoulidis K, Ivanov I, Anghel Drugarin C V 2018 Electric power losses of frequency controlled electric drive with high-voltage induction motor, *International Conference on Applied and Theoretical Electricity 2018 (ICATE 2018)* Craiova Romania (IEEE *Xplore*® Catalog Number: CFP1899S-ART)

[9] Chapman S J 1985 *Electric Machinery Fundamentals* (McGraw-Hill, N. Y.)

[10] Kluchev V I 1989 *Electric Drive Theory* (Tehnika, Sofia) (in Bulgarian)

[11] Rachev S, Koeva D. (eds) 2016 Electromechanical model and operating modes of high-voltage induction motor electric drive – part I: impact torque and currents *International Symposium on Fundamentals of Electrical Engineering (ISFEE)*, University Politehnica of Bucharest, Romania

[12] Crowder R 2006 *Electric Drives and Electromechanical Systems* Oxford UK pp 35-42

[13] Rizzoni G 1993 *Principles and Applications of Electrical Engineering* (Illinois, USA Richard D. Irwin) pp 853–855

[14] https://www.elprommotors.com

Acknowledgements

The implementation of this paper is thanks to the support from:

- the University Project 2104E 'Electric drives for electric vehicles and industrial applications, electrical components and systems – practical and mathematical model studies in terms of energy and economic efficiency', financed by the Ministry of Education and Science of the Republic of Bulgaria
- the Project BG05M2OP001-1.002-0023 Competence Center 'Intelligent Mechatronic, Eco- and Energy-Saving Systems and Technologies' financed by the Operational Program 'Science and Education for Intelligent Growth' of the European Regional Development Fund.

IOP Conf. Series: Materials Science and Engineering 1216 (2022) 012008 doi:10.1088/1757-899X/1216/1/012008

Statistical approach and multiplicative models for electric vehicles charging behaviour patterns

D Y Koeva[1]

[1] Department of Electrical Engineering and Electronics, Technical University of Gabrovo, 4 H Dimitar, Gabrovo 5300, Bulgaria

E-mail: dkoeva@abv.bg

Abstract. Since the charging processes of electric vehicles are stochastic and time-dependent, the paper views an approach based on a statistical analysis of real data on electricity consumption at charging station connection points. Other types of data (geographical, public sites, distance between individual charging stations, etc.) are also taken into account when making the analysis. Multiplicative models are the most suitable for studying and forecasting time series with pronounced cyclicity and seasonality. Their application allows us to consider the correlation of the load in the consuming nodes with regional features, climatic factors and seasonality. The method and approach discussed in this paper make possible the processing of a large amount of data and the detection of load cyclicity in the load schedule of electricity facilities. The results of the model will identify the requested charging power in a developing charging infrastructure.

1. Introduction

The ever-increasing number of electric vehicles and the consequent major changes in transport sector that we are witnessing today pose enormous challenges to the electricity distribution network and the charging infrastructure connected to it. A careful stakeholder analysis is needed for the proper expansion of the charging infrastructure: from the point of view of electricity suppliers (appropriate place to provide the necessary charging power without peak overloads of electricity consumption at the point of connection of the charging station); from the point of view of owners of electric vehicles (utilization of the structure for maximum accessibility and services); from the point of view of government and local government structures. Nowadays the remarkable evolution in the field of mobility is related to digitalization, autonomous driving, shared mobility and dynamically evolving electric mobility.

The reduction of CO2 emissions, atmospheric pollution and noise and the controlled dependence on oil are at the focus of this sustainable development. But how do we evaluate the participation of electric vehicles (EVs) in the implementation of each of these important goals? It seems that they are quite ambitious and the current 5.3 million (according to the International Energy Agency) electric vehicles will not cover them. Therefore, there is an urgent need to improve charging technology and build relevant charging station infrastructure. The charging process of electric vehicles is stochastic and time-dependent and the most common practice is to connect charging stations to an existing electricity supply network. The energy system operators are responsible for managing energy flows and constantly monitoring the balancing energy at the connection points of large consumers with peak loads, such as charging stations (CS). There are two main reasons why they do this: the load in the local distribution networks and the required transmission network are strongly increasing with all the negative consequences of this. The energy market regulation is a constant corrective. In Bulgaria, the development of the charging station infrastructure (CSI) turns out to be a simple function of the number of registered electric vehicles, without any research or analysis done in advance. There is no profile that prevails at the moment – business customers, as well as private and municipal installations are approximately equal. It is important to expand the network, first at public places and then at company parking lots. Business desire depends on the number of installed CSs and they are the most difficult thing in the field of electric mobility since more preparation is required. The secondary market for electric vehicles is developing slowly, mostly because of the undeveloped and incomplete CSI There are several web sites providing information about charging stations in Bulgaria, as well as worldwide. However, there is not a real platform for charging infrastructure management with various options for planning, booking, payment and management, as well as for attracting other participants in this process. In conclusion, the main steps towards low-carbon economy and more sustainable transport with the increasing use of EVs are careful forecast analysis of appropriate multiplicative models for electricity consumption and subsequent integration of renewable energy sources in the charging infrastructure.

2. Background

The structure of the energy distribution network to which charging stations are connected is interdependent and therefore a careful preventive analysis is recommended in order to avoid peak loads. Some factors that determine the behavior during charging are time-dependent, others are geographical and local, but overall the charging process is defined as stochastic. The results of an optimized multiplicative model for electricity consumption at the connection point of the charging station will be an indication of the requested power and location of the expanding charging infrastructure. This is particularly true for countries like Bulgaria, where the rate of use of EVs is described as low, „exploratory", and charging infrastructure as slowly developing and chaotically positioned.

The uncontrolled charging combined with the strongly increasing trend of EVs cause major problems in the distribution networks at some places. The solution is „smart charging management algorithm strategy" based on multiplicative forecasting models. In [1, 2] it is shown how the power transformers in PS overload when CS are connected. In [1, 2, 3], stochastic models based on the behavior of EVs' owners during charging are offered. Selected models are used to evaluate the impact of CS on the distribution network, as well as a tool to compare different types of EVs and the consumption of residential sites and other objects nearby. Another stochastic model is used in [4, 5] and it is based on the consumption of households supplied by the urban distribution network. Since 2016 the analyses of potential risks of charging according to geographical region, travel time between work and home, average mileage, charging time, etc. have started. This is how the so-called „charging profile of a typical EV battery" is built. Nowadays, according to [5], it is evident that studying travel patterns and charging behavior is beneficial.

The first approach involves the collection and processing of data for the purpose of forecast analysis of the required charging power.

The second approach is based on building a model of the entire EVs lot and using the data as an indicator of the required power provided by the charging infrastructure at certain points and times, [6]. Obviously, the forecast analysis of the energy consumed by the charging stations is extremely important when it comes to evaluating the load on the distribution networks and determining the exact location and power when the CSI is expanding.

3. General statement and approach

The most common models for identifying discrete objects and time-dependent processes are:

3.1. Autoregressive model – AR (Auto Regressive)

It is considered as the simplest and described as follows:

$$A(z).y(t) = \mathrm{e}(t) \tag{1}$$

$$A(z) = 1 + a_1.z^{-1} + a_2.z^{-2} + \dots + a_{n_a}.z^{-n_a} \tag{2}$$

3.2. ARX model – (Auto Regressive with eXternal input)
In generalized and expanded form is given with the expressions:

$$A(z).y(t) = B(z).u(t) + e(t) \tag{3}$$

$$y(t) + a_1.y(t-1) + \dots + a_{n_a}.y(t-n) = b_1.u(t) + b_2.u(t-1) + \dots + b_{n_b}.u(t-m) + e(t) \tag{4}$$

$$B(z) = b_1 + b_2.z^{-1} + \dots + b_{n_b}.z^{-n_b+1} \tag{5}$$

Optimization with ARX models is most effective when applying a polynomial approximation. The forecasts made by ARX models have a high degree of accuracy, especially if the approximate polynomials are of a higher order. Distortions (disturbances) as part of the dynamic system are a disadvantage. These models are widely used because of their main advantage: it is not necessary to know the physical characteristics of the system, which greatly reduces engineering work.

3.3. ARMAX model – (Auto Regressive – Moving Average with eXternal input)
In this case, the object of study (electricity consumption) is time dependent. The models used for forecasting are non-deterministic as it is assumed that they allow us to determine the probability of future values in a certain range. Such models are called stochastic and they play an important role. They are divided into stationary and non-stationary. Stationary models are based on the assumption that the process remains in equilibrium around a relatively constant average level. But often in industry, trade, economy and so on, there are time series that are best described as non-stationary, which means that they do not have a natural average level. In this case an exponentially weighted moving average value is used. The stochastic model, where forecasting by exponentially weighted moving average is optimal, belongs to the class of non-stationary processes, also known as autoregressive processes - with integrated moving average or ARIMA (Auto Regressive Integrated Moving Average) models.

$$A(z).y(t) = B(z).u(t-n_k) + C(z).e(t) \tag{6}$$

where n_k is a value that characterizes the delay.

$$C(z) = 1 + c_1.z^{-1} + c_2.z^{-2} + \dots + c_{n_c}.z^{-n_c} \tag{7}$$

To achieve greater flexibility in the construction and operation of the model in terms of the observed time series, sometimes it is more appropriate for the models to be combined in one autoregression and moving average model. The combined model is described by the expression:

$$\widetilde{z_t} = \phi_1.\widetilde{z_{t-1}} + \dots + \phi_p.\widetilde{z_{t-p}} + a_t - \theta_1.a_{t-1} - \dots - \theta_q.a_{t-q} \tag{8}$$

where p+q+2 are unknown parameters: μ, $\phi_1, \phi_2, \dots, \phi_p$, $\theta_1, \theta_2, \dots, \theta_q$, σ_a^2, which are determined by the observations.

It is important to know that the combined models have the following distinguishing features: stationarity, reversibility and reciprocity. Mixed autoregression – the moving average model is always preferred because of its economy. In general, the process is expressed by the equation:

$$\widetilde{z_t} = \phi_1.\widetilde{z_{t-1}} + \dots + \phi_p.\widetilde{z_{t-p}} + a_t - \theta_1.a_{t-1} - \dots - \theta_q.a_{t-q} \tag{9}$$

$$\left(1-\phi_1.B-\phi_2.B^2-\cdots-\phi_p.B^p\right).\tilde{z}_t=\left(1-\theta_1.B-\theta_2.B^2-\cdots-\theta_q.B^q\right).a_t \quad (10)$$

$$\phi(B).\tilde{z}_t=\theta(B).a_t \quad (11)$$

where: $\phi(B)$ и $\theta(B)$ - are polynomials of B of degree p and q.

Many time series in industry, trade, economy are non-stationary and do not fluctuate around a relatively fixed average value. Along with that, the process fluctuates around a different average level and, taking into account the differences in levels, a similar behavior of the order is observed. Such an order can be represented by a generalized autoregression operator φ (B) where one or more zeros of the polynomial φ (B) are equal to 1. When similar features of the time series are repeated after the s interval (they are called reference intervals), it is said that the time series has periodicity s. The most common mistake in the analysis of time series is mixing the procedures for building the forecasting model. Therefore, a common approach to make the analysis is to divide the order into three components: trend, seasonal component, and random component. The multiplicative models are suitable for researching and forecasting time series with pronounced seasonal periodicity. If the time series is $\{Z_t, t=1,2,3,\dots,k\}$, a SARIMA model (Seasonal ARIMA model) is generated for it from row (p, d, q) x (P, D, Q) with average value μ, as follows [6, 7, 8, 9]:

$$\phi_p.\Phi_P(B^S).(1-B)^d.(1-B^S)^D.(Z_t-\mu)=\theta_q(B).\Theta_Q.(B^S).a_t \quad (12)$$

where: $\phi(B)$ is a polynomial of order *p*, *p* is the order of non-seasonal autoregression; $\theta(B)$ - polynomial of order *q*, *q* - the order of the non-seasonal component of the moving average; $\Phi(B^S)$- polynomial of order *P*, *P* - order of seasonal autoregression; $\Theta(B^S)$ - polynomial of order *Q*, *Q* - the order of the seasonal component of the moving average; *d* - the number of differences; *D* - the number of seasonal differences; *S* - the length of the season.

The development and analysis of the model is performed in the following sequence:

Descriptive Methods – the data are visualized and the frequency of seasonal occurrence is established, if it is not precisely determined in advance;

Seasonal Decomposition – the data are divided into four components: trend (T_t), cyclical component (C_t), seasonal component (S_t), and random component (R_t). The data are presented as a product of the individual components $Y_t=T_t.C_t.S_t.R_t$;

User-Specified Model – the most appropriate models are selected;

Automatic Forecasting – automated model selection is performed with the best approximation to the source data.

Estimated Autocorrelations, Partial Autocorrelations and Tests for Randomness of residuals – the autocorrelation function, the partial autocorrelation function and a randomness test of the difference between data and forecast are calculated, i.e. the residual error. The type and values of the functions are used to diagnose the model.

4. Results and discussion

The object of monitoring and analysis is the consumption of three publicly available charging stations (CS 1, CS 2 and CS 3), connected to the city distribution network. Considering the problems described in sections 1 and 2, we use the data collected for the energy profiles of the three power substations (PS 1, PS 2, PS 3) supplying these terminals, Figure 1. The data include active and reactive electricity measured in 15 minutes for three typical months: January, May and July, (2976 values for each month), 2020. The energy profile of the terminals to which the CSs are connected is monitored at the same time, Figure 2, Figure 3, Figure 4. The energy profiles represent active and reactive energy for one typical working day of each month, a total of 100 measured values for 24 hours (x-axis). The analysis of the application results and the optimized multiplicative model will be an indication of the correlation of the load with climatic factors, seasonality, electricity tariff zones and EVs charging behavior.

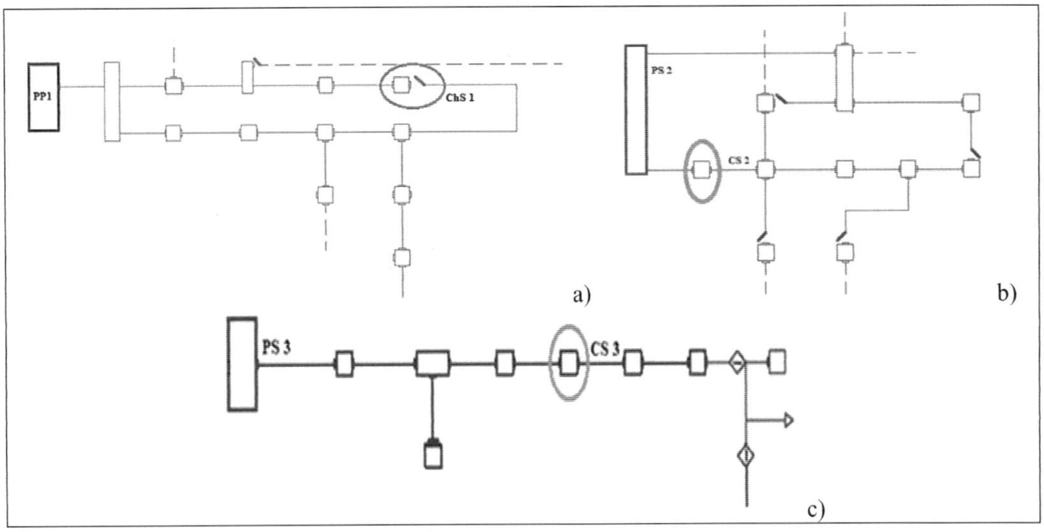

Figure 1. Power substations PS and CSs location: PS1 and CS1 (a), PS2 and CS2 (b), PS3 and CS3 (c).

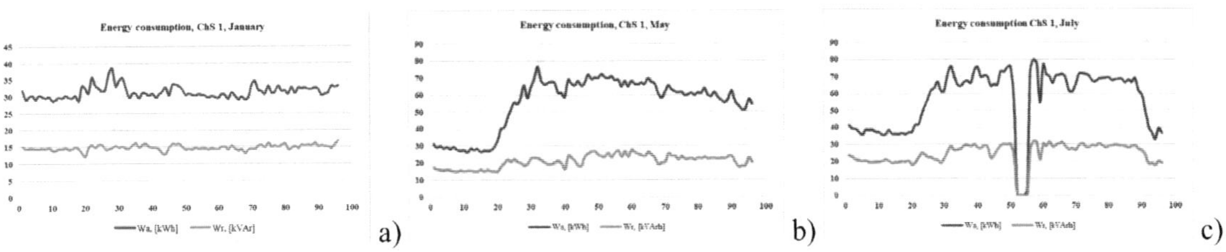

Figure 2. CS 1 energy profile for January (a), May (b) and July (c), 2020.

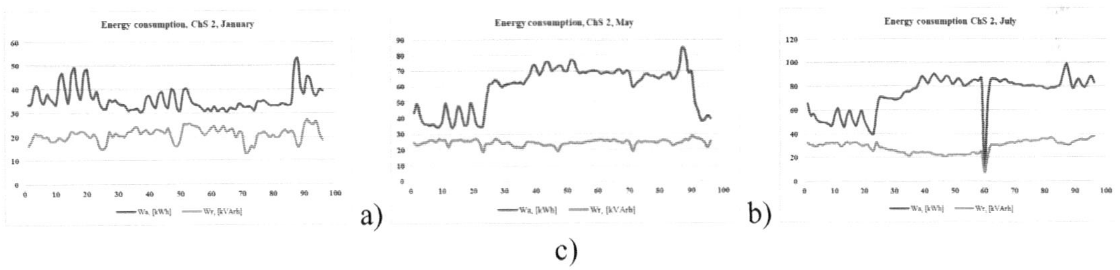

Figure 3. CS 2 energy profile for January (a), May (b) and July (c), 2020.

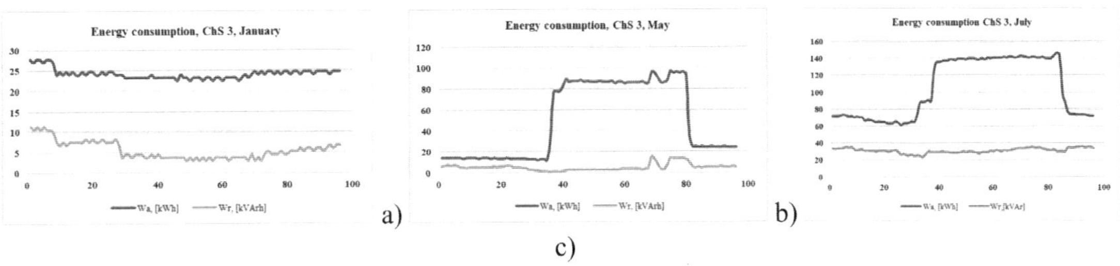

Figure 4. CS 3 energy profile for January, May and July 2020.

Table 1. Initial data processing.

	Unique Values	Minimum	Maximum
CS1	57	12	146.25
CS2	116	10.2	98.8
CS3	136	0	79.8

Table 2. ARIMA Model Summary

Parameter	Estimate	Standard Error	*t*	P- value
AR (1)	0.885454	0.0192577	45.9793	0
MA (1)	-0.18089	0.0397865	-4.54652	5E-6
SMA (1)	0.646419	0.0289428	22.3343	0

In a three-coordinate system, the data from January (x-axis), May (y-axis) and July (z-axis) are compared, the so-called data distribution density, Figure 5 a). Each monthly energy profile is checked for random data (Normal Probability plot with 95% limits). Box and Whisker plot helps to find datacenter, median and interquartile range, as it is done for May, figure 6. Because of the processing of 288 data on the consumption of CS for each month considered, non-repeating, minimum and maximum values for each of them are established. Table 1. The density of distribution is presented on figure 5 b). A completely adequate description of the seasonal time series using a multiplicative model is impossible to obtain. The adequacy of the model gives an idea about how accurate (as close as possible to the specific data) and reliable it is (in which cases it gives good results).

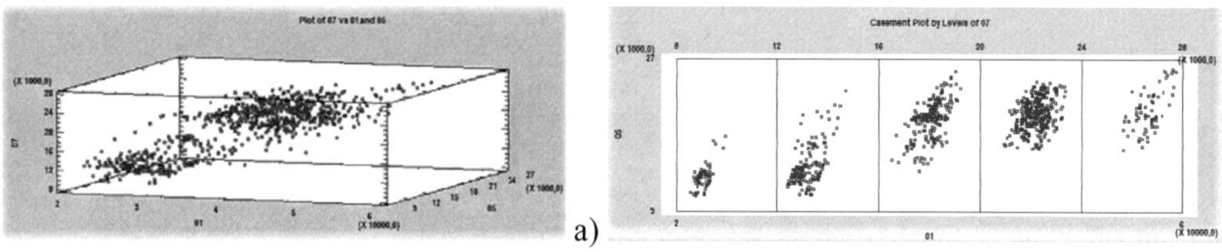

Figure 5. The density of distribution.

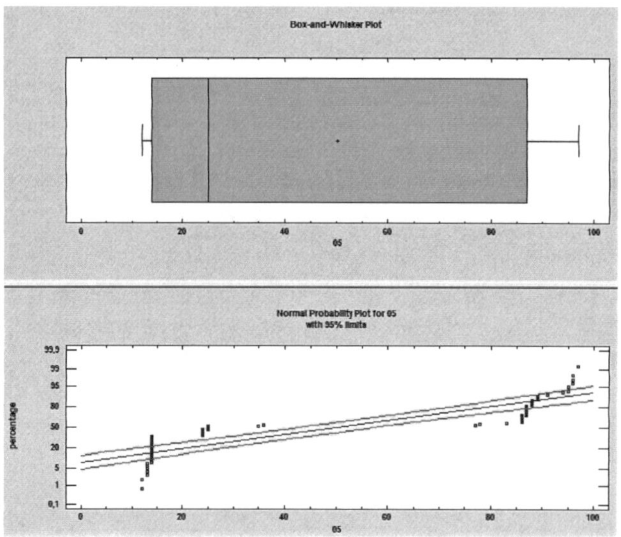

Figure 6. Data randomness check.

The adequacy analysis is performed after calculating the residual error once the model has been started. An autocorrelation checks of the residual errors using the method of least squares is made. In addition, a check of a periodogram that shows the probabilistic distribution of the residual error is made. However, both checks are considered relatively rough and do not provide an accurate picture of the adequacy of the model. Selected from a total of seven models, the ARIMA $(1,0,1)x(0,1,1)_7$ model for forecasting the daily profile of electricity consumption gives very good results, Table 2. According to the analysis of the coefficients of the autocorrelation and the private autocorrelation functions of the residue, only between two private autocorrelation functions out of 24 been calculated a correlation is found. The forecast can be more precise if the relationship between the residual of the data after the forecast analysis and the predicted values is used, Figure 7:

$$R = e^{[0,282929+0,980795.\ln(F)]} \tag{13}$$

where: R – Residual; F – Forecast.

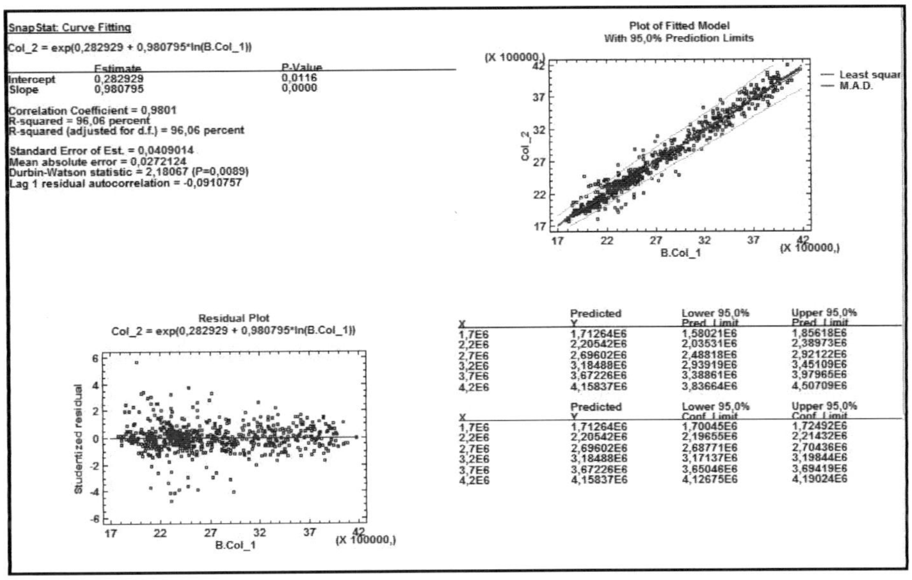

Figure 7. Finding a relation between predicted values and residual data.

5. Conclusion

The disadvantage of the model is the difficulty in predicting values with a low frequency of occurrence, but in all cases, the quality of the data used for the model is of great importance. It is necessary to have a long period of observations and thus to form a database of values of the observed parameters and their change. The electricity consumed in January differs from the trend and cyclicity of those in May and July. The reason is compliance with all restrictions and lockdown due to COVID-19. These measures affect households and industry, and therefore the whole economy. Therefore, another statistical approach is required to study their impact.

6. References

[1] Brady J, O`Mahony M 2016 Modelling charging profiles of electric vehicles based on real-world electric vehicle charging data *Sustainable Cities and Society* vol 26 p 203-216 doi: 10.1-16/ j.scs.2016.06.014

[2] Huang Y 2019 Day-Ahead Optimal Control of PEV Battery Storage Devices Taking Into Account the Voltage Regulation of the Residential Power Grid *IEEE Transactions on Power Systems* vol 3, no 6, doi: 10.1109/TPWRS.2019.2917009

[3] Muratori M, Morgan M J, Serra E, Rizzoni G 2013 Highly resolved modeling of personal transportation energy consumption in the United States *Energy* vol 58 pp 168-177

[4] McKenna E, Thomson M 2016 High resolution stochastic integrated thermal-electrical domestic demand model *Applied Energy* vol 165 pp 445-61

[5] Arghira N, Hawarah L, Ploix S, Jacomino M 2012 Prediction of appliances energy use in smart homes *Energy* vol 48, no 1 pp 128-134

[6] Pevec D, Babic J, Kayser M, Carvalho A, Ghiassi-Farrokhfal Y, Podobnic V 2018 A data-driven statictical approach for extending electric vehicle charging infrastructure *Energy Research* vol. 42 pp 3102-20 https://doi.org/10.1002/er.3978

[7] Dong Y, Wang J, Jiang H, Wu J 2011 Short-term electricity price forecast based on the improved hybrid model, *Energy Conversion and Management* vol 52 pp 2987-2995 doi: 10.1016 /j.enconman.2011.04.020

[8] Wang Sh, Feng J, Liu G 2011 Application of seasonal time series model in the precipitation forecast, *Mathematical and Computer Modelling* vol 58, pp 677-683 doi: 10.1016/ j.mcm.2011.10.034

[9] Box G E P, Jenkins G M, Reinsel G, Ljung G 2015 *Time Series Analysis: Forecasting and Control* 5th Edition, Holden-Day San Francisco ISBN: 978-1-118-67502-1

Acknowledgements
The implementation of this paper is thanks to the support from the University Project 2104E "Electric drives for electric vehicles and industrial applications, electrical components and systems – practical and mathematical model studies in terms of energy and economic efficiency", financed by the Ministry of Education and Science of the Republic of Bulgaria.

A mathematical model for the virtual simulator of the power unit electrical part

P Baran, Y Varetsky, V Kidyba and Y Pryshliak

Department of Power Engineering and Control Systems, Lviv Polytechnic National University, Lviv 79013, Ukraine

E-mail: j.varetsky@gmail.com

Abstract. The mathematical model is developed for a virtual training system (simulator) of the power unit electrical part operators of a thermal (nuclear) power plant. The model is used to simulating the main operating conditions of the power unit electrical part: generator idling, generator synchronization with the power system, excitation shifting from the main unit to the backup one and vice versa, switching in the power unit auxiliary system, and others. Furthermore, it has been implemented modelling some probable emergency conditions within a power plant: incomplete phase switching, damage to standard power unit equipment, synchronous oscillations, asynchronous mode, etc. The model of the power unit electrical part consists of two interacting software units: models of power equipment (turbine, generator with excitation systems, auxiliary system) and models of its control systems, automation, relay protection and signalling. The models are represented by the corresponding algebraic-differential equations that provide real-time mapping power unit processes at the operator's request. The developed model uses optimal solving algebraic-differential equations to ensure the virtual process behaviour in real-time. In particular, the implicit Euler method is used to solve differential equations, which is stable when simulating processes in significant disturbances, such as accidental disconnection of the unit from the power system, tripping and energizing loads, generator excitation loss, etc.

1. Introduction

Safe, reliable and efficient operation of a power plant can only be achieved with qualified service personnel. Their theoretical knowledge and practical experience assess the qualifications of the personnel. The operator experience allows avoiding mistakes and wrong decisions in planned or emergency scenarios of the power plant operation, excluding the possible production losses, equipment damage and even the power system blackout. Therefore, the operator must determine the correct sequence of actions in various operating situations. Suppose the operator did not faced some possible situations during the station operation. In that case, a virtual simulator can design them artificially in real time and display the process on the mnemonic control panel of the power plant or power unit. To this aim, various training systems have been created, ranging from the simplest, which studies the technical characteristics of power plant equipment, the basics of power unit management and improved training systems that reflect the behaviour of power units during operational management. The latest training systems, the so-called full-scale simulators or high-fidelity power plant simulators, utilize real-time scenarios as in an actual power plant [1, 2]. These simulators use powerful servers and can be adapted to create different challenges during actual power plant operations. The cost of development, creation and operation of such complexes reaches several million dollars [1].

It is possible to improve the quality of training of operational personnel of power plants due to the use of specialized (local) training systems [3]. These systems allow you to practice operational management of individual systems of the power unit, such as a steam generator (or its particular elements), a nuclear reactor, turbine, generator, etc. [4, 5].

2. Virtual training simulator outline

The authors have developed a virtual training simulator (VTS) to train operational personnel to control the electrical part of the power unit of both thermal and nuclear power plants [6]. Using a personal computer (PC) for the VTS operation essentially reduces its cost. In the presented system, in contrast to the full-scale simulator, there are no special technical means (control keys, switches, control devices, etc.) that reproduced the instruments of the control, monitoring and signalling of an actual power plant unit. The absence of these traditional control and monitoring bodies is not an essential shortcoming of the proposed system because PC control is widely used at modern power plants.

The developed VTS consists of two subsystems. The first subsystem carries out modelling technological processes in the power unit electric part. Such modelling should reproduce the unit electrical part behaviour in real-time at long-lasting intervals. For operational control, virtual models of panels and motor fields have been created. The VTS allows training the managing the electrical part of the unit during normal, abnormal and emergency operating conditions: generator idling, generator synchronization with the power system (manual, automatic), control of active and reactive power during unit operation, excitation shifting from the primary unit to the backup one and vice versa, shifting power supply of auxiliaries bus sections from the main transformer to the backup and vice versa, partial and complete loss of the generator excitation, power system synchronous oscillations, load asymmetry, etc.

The training subsystem implements the following operations: auto-simulator, demonstration, self-training, self-checking.

Auto-simulator mode is an operation without a scenario under the initial conditions defined by the selected exercise. The operator carries out any free actions, limited only by the technology requirements of the power unit electrical part (parametric restrictions, operation of relay protection devices and automation, etc.).

Demonstration mode is a step-by-step implementation of the selected exercise. All actions to control the unit electrical part are performed automatically; the operator only contemplates the consequences of acting.

Self-training mode is an operation by the scenario without a time limit with the indication of mistakes and the ability to seek help from the training subsystem.

Self-checking mode is an operation by the time-limited scenario, analyzing the number and importance of mistakes with quantitative assessment of the trainee for a given exercise.

Scenarios represent a sequence of actions that the operator must perform during the execution of a given exercise. A system of encoding and decoding actions to perform a given exercise has been developed to generate scenarios.

3. The mathematical model for virtual simulator

The mathematical model is developed for a virtual training system (simulator) of the power unit electrical part operators of a thermal (nuclear) power plant. The model of the power unit electrical part consists of two interacting software units: models of power equipment (turbine, generator with excitation systems, auxiliary system) and models of its control systems, automation, relay protection and signalling. The models are represented by the corresponding algebraic-differential equations that provide real-time mapping power unit processes at the operator's request. The developed model uses optimal methods for solving algebraic-differential equations to ensure the virtual process behaviour in real-time. In particular, the implicit Euler method is used to solve differential equations, which is stable when simulating processes in significant disturbances, such as accidental disconnection of the unit from the power system, tripping and energizing loads, generator excitation loss, etc.

3.1. Turbine model

A simplified turbine model with automatic speed control has been developed to simulate the generator idling, its synchronization with the power system, active power control during regular operation. The model is written for per unit variables as follows:

$$T_P\cdot\frac{d\mu_P}{dt}=\begin{cases}0, & \mathrm{mod}(\psi-\mu_P), \le b_{ZH};\\ \mathrm{sign}(\psi-\mu_P-b_{ZH})\cdot\mathrm{mod}(\psi-\mu_P); & \end{cases}$$

$$\psi=\frac{(\omega_0-\omega_G)}{\delta}; \tag{1}$$

$$M_T=\mu_P,$$

where T_P, b_{ZH}, δ is time constant, characteristic of the insensitivity and automatic speed controller slop, respectively; μ_p is rod move of the servomotor controlling the turbine control valves, in per units; ω_0 is frequency setpoint for the automatic speed controller; ω_G is turbogenerator rotor speed; ψ is a position of the centrifugal pendulum clutch for automatic speed controller; M_T is the turbine torque.

3.2. Turbogenerator electromagnetic model

The turbogenerator model implements Park equations of the synchronous machine in d, q coordinates. The model accounts for rotation e.m.f., without the impact of transformation e.m.f. and equivalent damping circuits on the generator stator voltage because they present only in transients. Rotor damping circuits is used in the generator rotor swing equations. Therefore, the electromagnetic model of the turbogenerator is as follows:

$$\begin{vmatrix}U_d\\ -U_f\\ U_q\end{vmatrix}=\begin{vmatrix}r & 0 & \omega_G\cdot L_q\\ 0 & r_f & 0\\ -\omega_G\cdot L_d & -\sqrt{3/2}\cdot\omega_G\cdot L_{ad} & r\end{vmatrix}\times\begin{vmatrix}I_d\\ I_f\\ I_q\end{vmatrix}, \tag{2}$$

where U_d, U_q, I_d, I_q is components of turbogenerator stator voltage and current vectors for the d and q axis; U_f, I_f is excitation winding voltage and current; r, r_f is resistances of the stator and excitation windings; L_d , L_q is d and q stator winding inductance components; L_{ad} is the mutual inductance between the stator and the excitation windings for the d axis.

3.3. Turbogenerator electromechanical model

Changing rotation speed of the turbine-generator shaft accompanies electromechanical operating conditions (for example, the unit idling, asynchronous mode, power unit synchronization). To modelling those conditions, the turbogenerator swing equation supplements the VTS mathematical model:

$$T_J\cdot\frac{d\omega_G}{dt}=M_T-M_{EM}-M_0, \tag{3}$$

where T_J is the inertia constant of the power unit; M_T is turbine torque; M_{EM} is electromagnetic torque; M_0 is the mechanical loss torque.

The electromagnetic torque in the per units considering equal inductances along the d and q axis is as follows:

$$M_{EM}=\frac{1}{3}\cdot\left(L_{ad}\cdot I_q\cdot(I_d+I_f)-L_{ad}\cdot I_d\cdot I_q\right)+M_D\cdot\omega_s, \tag{4}$$

where M_D is a damping factor; ω_s is the slip frequency.

The mechanical loss torque accounts for the purely mechanical power loss due to friction in the bearings:

$$M_0=a_1\cdot\omega_G+a_2\cdot\omega_G^2+\frac{a_3}{a_4+a_5\cdot\omega_G}, \quad (5)$$

where the coefficients a_1, a_2, a_3, a_4, a_5 are constant for the specific turbine.

The following equations relate the generator stator d, q voltage components and the power system voltage:

$$U_d=U_c\cdot\sin\delta+x_c\cdot I_{cq};$$
$$U_q=U_c\cdot\cos\delta+x_c\cdot I_{cd}, \quad (6)$$

where U_c is the power system voltage reduced to the generator voltage; x_c is the power system reactance; δ is the rotor angle; I_{cq}, I_{cd} is longitudinal and transverse current components supplied by to the system, and

$$I_{cd}=I_d - I_{dBB};$$
$$I_{cq}=I_q - I_{qBB}, \quad (7)$$

where I_{dBB}, I_{qBB} – power unit auxiliaries current components.

When supplying power unit auxiliaries from an auxiliary transformer, the current of the transformer is determined within the unit auxiliary model. In the case of connecting electric drives of the unit auxiliary to the standby transformer, I_{dBB}, I_{qBB} are assumed to be equal to zero. Equations (6) and (7) are supplemented by an equation that relates the generator angular frequency and the system angular frequency:

$$\frac{1}{\omega_{nom}}\cdot\frac{d\delta}{dt}=\omega_G - \omega_c, \quad (8)$$

where ω_{nom} is the nominal angular frequency, ω_c is the system angular frequency.

3.4. Turbogenerator excitation system model

Modern power units use, as a rule, thyristor or brushless excitation systems for powering the generator field winding. The automatic control of the excitation systems is equipped with power system stabilizers (PSS). The model of such an excitation system is presented as follows:

$$e = k_{0U}\cdot(U_G - U_0) + k_{1U}\cdot\frac{dU_G}{dt} + k_{1f}\cdot\frac{dI_f}{dt};$$
$$e_{min}\le e\le e_{max}; \quad (9)$$
$$U_f = e,$$

where e is the signal at the output of the controller; U_0 is controller voltage setting of the generator; k_{0U}, k_{1U}, k_{1f} are gains of the corresponding controller channels, e_{min}, e_{max} are the controller output signal limits.

Because the power system within the model is presented as a source of constant frequency and voltage, the model of the controller has no channels for frequency deviation and frequency derivative. The controller model can perform several other functions: excitation forcing, rotor current limiting when forcing at double level, automatic generator unloading when excitation overcurrent, automatic setpoint change during synchronization, limiting field current when reactive power consumption, and generator overvoltage protection at unexpected load tripping.

3.5. Auxiliaries model

The main components of the power unit electrical part are powerful induction motors (IM). The following equation presents the IM rotor electromechanical motion:

$$F(s_i, t)=T_{ji}\cdot\frac{ds_i}{dt} - M_{EMi} + M_{Ci}=0, \tag{10}$$

where T_{ji} is the mechanical time constant of the i-th drive; M_{Ci} is a mechanical moment of this drive; M_{EMi} is the motor shaft electromagnetic torque; s_i is the i-th IM rotor slip.

In the general case, a mechanical moment of a drive is a nonlinear dependence on the IM rotor rotation speed [7]. In the elaborated model, the mechanical moment is determined by a function:

$$M_{Ci}=k_{zi}\cdot\left(M_{CTi} + (1 - M_{CTi})\cdot\left(\frac{1-s_i}{1-s_{nomi}}\right)^{n_i}\right), \tag{11}$$

where k_{zi} is the degree of the i-th mechanism loading; M_{CTi} is the initial mechanical moment of the i-th drive (at slip frequency $s_i= 1$); s_{nomi} is the IM_i nominal slip; n_i is the i-th mechanism type exponent (for mechanism with a non-slip dependence $n_i = 0$, for fan $n_i = 2$, for different pump types, the value of n_i varies between 2 – 4).

To the IM shaft electromagnetic torque model M_{EMi} we used a multi-contour electrical circuit, accounting for the rotor current skin effect [8].

3.6. Solving Model Equations

The developed model uses optimal methods to solving algebraic-differential equations ensuring the virtual process behaviour in real-time. In particular, the implicit Euler method is used to solve differential equations, which is stable when simulating processes in significant disturbances, such as accidental disconnection of the unit from the power system, tripping and energizing loads, generator excitation loss, etc.

The system of differential equations of the developed model in the implicit Euler method can be represented as:

$$\vec{y}_{k+1} - \vec{y}_k - h\cdot\vec{f}(\vec{y}_k, t)=0, \tag{12}$$

where $\vec{y}_{k+1}$ is a vector of the operating condition parameters at a given integration step; $\vec{y}_k$ is a vector of the operating condition parameters at the previous integration step; $\vec{f}(\vec{y}_k, t)$ is a vector-function of these operating condition parameters; h is an integration step.

The vector of the operating condition parameters for the mathematical model of the power unit electric part (without auxiliaries), concerning which the algebraic-differential equations is solved, can be presented as follows:

$$\vec{y}=\left(\mu_p, \omega_G, \delta, U_d, U_q, I_d, I_q, I_f, U_p\right). \tag{13}$$

Newton's method with the following iterative formula is used to solve the equations (11):

$$\begin{gathered} \breve{W}(\vec{f}(\vec{y}^{(l)}_{k+1}))\cdot\Delta\vec{y}^{(l)}_{k+1} = \vec{y}^{(l)}_{k+1} - \vec{y}_k - h\cdot\vec{f}(\vec{y}^{(l)}_{k+1}, t); \\ \vec{y}^{(l+1)}_{k+1} = \vec{y}^{(l)}_{k+1} - \Delta\vec{y}^{(l)}_{k+1}, \end{gathered} \tag{14}$$

where $\vec{y}^{(l)}_{k+1}$ is a vector of the operating condition parameters for the l - iteration at the $k+1$ integration step; $\breve{W}(\vec{f}(\vec{y}^{(l)}_{k+1}))$ ·is the Jacobi matrix of the equations (12).

The solution of the equations (14) is carried out by the LDU transformation algorithm based on the Gaussian method. According to this algorithm, the original system of equations is rearranged to obtain the minimum number of non-zero coefficients during the transformation of the Jacobi matrix; trivial operations are excluded (multiplication and addition of zero, multiplication by 1, etc.). In order to optimize the solution of the equations (14), the Jacobi matrix is written in tape form. This solution optimization is primarily aimed to achieve real-time simulations. Software implementation of digital models was carried out in the programming environment Lazarus [9]. The use of this software environment allows compiling a program to work in different operating systems.

4. Modelling virtual simulator interface

The power unit electrical part model is managed by the model of the power block control panel, which displays on the computer monitor a mnemonic control panel with monitoring and signalling devices. Based on object-oriented programming technologies has been created a specialized editor for designing a power unit control panel, including the location and appearance of control devices and switches, their connection to the power circuit, as well as the emergency and process signalling boards. Then the appearance of the managing, monitoring and signalling devices on the control panel model and their mutual placement corresponds as much as possible to the prototype of the actual control panel.

A blender package for 3D graphics modelling was used to create graphic images of control panel components [10]. To simulating the control, monitoring and signalling devices of the power unit electrical part control panel, a library of the model components has been created.

The use of cyclic rendering (Cycles Render) provided high-quality images of the model components on the LCD monitor, as close as possible to their actual appearance. To visualize the labels and device arrows, the additional module BGRABitmap was used, and antialiasing and transparency effects were applied, which significantly improved the realism of displaying control and monitoring devices on the LCD monitor.

The developed virtual training simulator for a 300 MW power unit of a thermal power plant was commissioned in the Ukrainian power system. Figure 1 depicts a display of the turbo generator TGV-300 control panel fragment on a liquid crystal PC monitor of the Virtual Training Simulator.

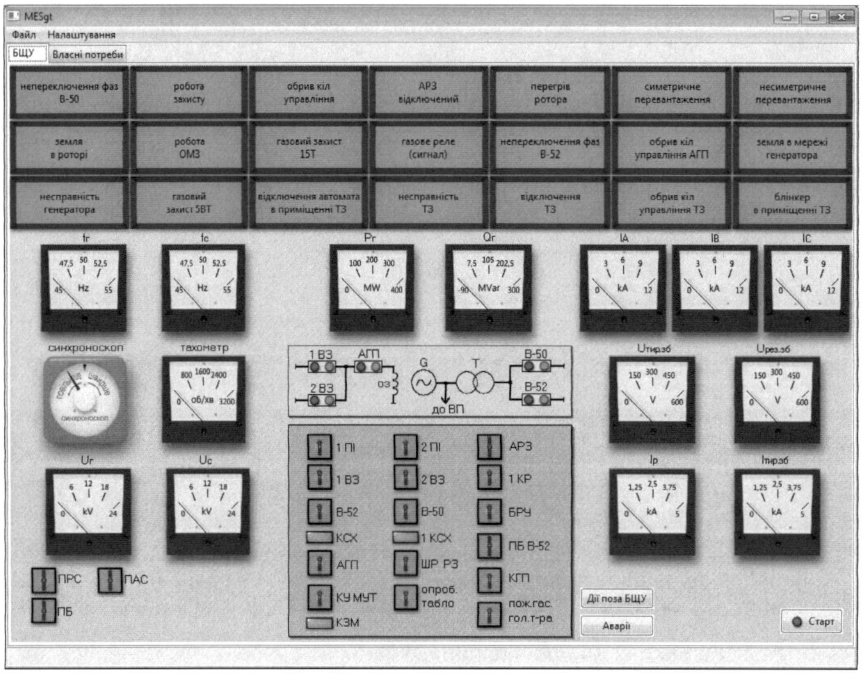

Figure 1. PC monitor display of the turbo generator TGV-300 control panel fragment.

5. Conclusions

The developed virtual training system (VTS) aims to train operational personnel to manage the power unit electrical part of a thermal or nuclear power plant. The mathematical model of the power unit electrical part, including its digital implementation, allows simulating electromechanical processes in real-time at long intervals with the required accuracy. The use of special programming tools has allowed obtaining high-quality graphic images of the control panel components on a PC monitor, enhancing the sense of reality when working with the virtual training system. The developed virtual training system can be used in the curriculum of power engineering students to gain knowledge in the field of operational management of power plants.

6. References

[1] Magid S I and Arkhipova E N 2017 *Training of energy personnel* (Moscow: Energobezopasnost Publishing House)

[2] Full Scope Simulators. https://www.l3harris.com/all-capabilities/full-scope-simulators

[3] Shcherbnev V S, Ilchenko A G and Tokov A Yu 2007 *Bulletin of the Ivanovo State Power Engineering University* **2** pp 40–42

[4] Complex exercise machines. https://www.asot.com.ua

[5] Simulators for power plants. http://fpps.ru/simulators

[6] Baran P M, Kidyba V P, Pryshliak Y D and Dembitsky M I 2015 *Bulletin of the Lviv Polytechnic National University. Power engineering and electromechanical systems* **834** pp 3–7

[7] Gurevich Yu E 2008 *Application of mathematical models of electric load in calculations of stability of power systems and reliability of power supply of industrial enterprises* (Moscow: Eleks-KM)

[8] Sivokobylenko V F and Vasilets S V 2014 *Scientific Bulletin of the National Mining University* **5** pp 76–82

[9] Lazarus homepage. https://www.lazarus-ide.org

[10] Home of the blender. https://www.blender.org

IOP Conf. Series: Materials Science and Engineering 1216 (2022) 012010 doi:10.1088/1757-899X/1216/1/012010

Modern materialistic understanding of the "energy" notion and the method of its systemic evaluation in chemicals

O S Sirotkin[1] and R O Sirotkin[1]

[1] School of Materials Science and Engineering, Kazan State Power Engineering University, 420066, Kazan, Krasnosel'skaya str., 51 Russia

E-mail: sirotkin-49@mail.ru

Abstract. It was shown that the traditional approaches to understanding the notion of "energy" as a work or a physical quantity are outdated. For example, R. Feynman noted that "today's physics does not know what energy is." Therefore, even now, some researchers believe that "by and large, the concept of energy… is artificial, because unlike matter, of which we can say that it exists, energy is the fruit of human thought." In contrast to these ideas, the authors showed that energy, like matter, objectively exists in various forms (energy continuum), which differ in structure, and is able to perform different types of work, to determine the forms of interaction and movement of matter in various material systems (substances, material bodies and megamaterial systems). A new scientific foundation for systematization and quantitative evaluation of energy characteristics of chemical compounds was proposed. It is based on a comprehensive assessment of contribution of chemical compounds' composition and chemical bond type in line with a chemical bond's unified model and the "System of chemical bonds and compounds" (SCBC). As a result of using this basic scientific innovation, the symbiosis of Mendeleev's periodic table of atoms (composition – property) and SCBC (composition - chemical bond and structure – property) was realized for the first time. The foundation was laid for creating a database on systemic digitalization and evaluation of energy stored in various chemicals (natural gas, coal, oil, peat, wood, etc.) and the most effective ways of extracting it from these.

1. Introduction

The crisis in the evolution of scientific knowledge of the XXI century is becoming more and more obvious today. This is due to the presence of two major global *problems* in the development of scientific knowledge today. *First, the problems of scientific knowledge fragmentation and the development of a single universal view of the structure of the universe, as well as the clarification of the fundamental causes of the diversity of phenomena and objects of the Earth and the Universe as a whole, have become particularly acute in science, education and natural science in general. The second is the growing contradiction between the accumulated separate empirical information ("scattered" in around 17,000 different scientific disciplines) and a significant part of the traditional, but outdated scientific paradigms* [1-3]. That is, knowledge today is fragmented like a tree into separate branches, but the roots and the single unifying material trunk are not visible. Progress in the development of modern scientific approaches to the study of matter, its varieties, and the structure of the universe as a whole has also slowed down, in part due to attempts to equate scientific and materialistic approaches with mythology and religion (for example, in theology). However, the difference is obvious and these approaches are fundamentally opposite. Scientific materialism is based on the principle of the knowability of the universe (nature), while mythology is based on agnosticism, that is, on its unknowability. The reason for this difference is demonstrated by the radically different answer to the question: "What is primary: matter or consciousness?". For scientific materialism, matter is primary. Consciousness is secondary because it is only a property of matter, but not any matter, not stagnant and not any living thing, but only having a highly developed (as in an educated person) material organ – the brain. Today it is necessary to raise materialistic knowledge to a new integral level, taking into account the individual achievements of modern science in various (differential component) branches of knowledge, as well as educating our schoolchildren and students.

For power engineering universities, the concept of "energy" is of particular importance, but the meaning of its traditional definitions needs to be clarified and evolutionarily developed. However, the meaning of the concept of "energy" is traditionally presented not as an objective material reality, but only as a hypothetical physical quantity that can perform work, etc., that is, as a "property", but it is not clear whose property.

The aim of this work is to consider and clarify the meaning of the concept of "energy" in the framework of modern views on structure and properties of matter as the most important fundamental concept of natural science and to consider the option of a systemic assessment of chemical energy through a system of chemical bonds and compounds (SCBC) in the form of a "Chemical triangle".

2. Exposition

Historically, for a long time, instead of the term "energy", the term "living force" was used, introduced by I. Newton. For the first time in history of the concept of "living force", the meaning of "energy", without uttering this word, was put by Julius von Mayer in his paper "Remarks on the forces of inorganic nature", published in 1842. The special term "energy" was introduced in 1807 by the English physicist Thomas Young and denoted a quantity proportional to the mass and the square of the velocity of a moving body. In science, the term "energy" in its modern sense was introduced by William Thomson (Lord Kelvin) in 1860. The features of the properties of this phenomenon, as the "law of conservation and transformation of energy", was introduced into scientific circulation by F. Engels, this allowed all types of energy to be measured by using single unit.

In his lectures on physics, delivered in 1961-1963 at the California University of Technology, the famous American physicist R. Feynman noted [4] that within the framework of the law of conservation of energy "there is a certain quantity called energy, which does not change under any transformations occurring in nature." "This statement itself is very, very abstract; it is essentially a mathematical principle... "" It is important to understand that the physics of today does not know what energy is." Further in 2001 I. V. Plachkov et al. (Ukraine), when asked what the concept of "energy" is, noted that "by and large, the concept of energy, the idea of energy, is artificial and created specifically to be the result of our thoughts about the world around us." He further states, "that unlike matter, about which we can say that it exists, energy is the fruit of man's thought, his 'invention', constructed in such a way that it makes possible to describe various changes in the surrounding world and at the same time to talk about the permanence, the preservation of something that was called energy...".

It is hardly possible to agree with this, since *energy*, according to the authors of this paper, *is an objective* ***material*** *reality* that is widely used by humanity in solving problems of its life support (heat, electricity, etc.). And if so, then it is necessary to understand in what material form energy exists in the universe.

3. Discussion

The concept of "energy" is traditionally interpreted in different ways, including using traditional, but no longer accurate enough for the XXI century definitions. For example, *energy* (E) is a physical quantity that shows what work a body (or several bodies) can do, or the energy of a body. In another version, *energy* (Greek: ἐνέργεια - action, activity, force, power) is a scalar physical quantity that is a single measure of various forms of movement and interaction of matter, a measure of the transition of the movement of matter from one form to another. First, within the framework of the fourth (integral-differential) stage of the development of scientific knowledge and the paradigm of the multilevel organization of substance and matter [1-3], it is questionable whether it is correct to call any energy a "physical quantity" (after all, there are chemical and other types of energy). It is more correct to use a more precise and universal scientific concept of "material quantity". Secondly, energy is not just an abstract quantity that characterizes body's ability to work or a measure of various forms of motion and interaction of matter. These definitions are vulnerable due to the fact that "energy" in them is presented simply as a "property" of matter without revealing its material essence. After all, the property of energy is precisely the ability to bind discrete varieties of matter into various substances (atoms, molecules, cells) and material bodies, to produce work, etc. Therefore, for a consistent materialist, energy is an objective reality (thermal, electrical, nuclear, etc.) in the form of field forms (energy continuum) of matter and one of the two (together with mass) most important components and characteristics of material objects as a whole. It is this energy, together with mass, that determines, within the framework of the law of conservation of mass – energy and wave-particle dualism, the features of the modern paradigm of multi-level structural organization and properties of substances and matter in general (fields -substances - material bodies - megamaterial systems) [2, 3]. The contribution of the outstanding Russian scientist M. V. Lomonosov to the discovery of the law of conservation of mass is known. In 1905, the famous physicist A. Einstein showed that between the mass (m) of a body (substance) and its energy (E), there is the following relationship:

$$E = mc^2, \tag{1}$$

where c is the speed of light in a vacuum.

In this case, equation (1) is associated [5], not just with the energy E of any material body, but with the bond energy E_{bond} of elements in an atomic nucleus. Moreover, we usually consider the specific bond energy (the average bond energy per nucleon), which depends on the mass number A and the larger it is, the more stable the nucleus. For light nuclei (A is less than 12), with an increase in A, their E_{bond} increases sharply to 6-7 MeV. For example, for the heavy element isotope uranium-238, it reaches 7.6 MeV. For comparison, we note that the bond energy of valence electrons in atoms is 10 eV, that is, 10^6 times less than the nuclear energy. It is also worth noting that the mass of a nucleus is always less than the sum of the masses of its constituent nucleons (protons and neutrons), and this difference is called the *mass excess* Δm. This is due to the fact that during the nucleus formation, energy is released (part of the mass is spent on formation of a nuclear bond of elementary particles in a nucleus, characterized by the respective nuclear energy), and during the nucleus decay it must be spent. This effect demonstrates the relationship between the transformations of matter, accompanied by the *phenomenon (law) of conservation of mass-energy* and the transition of one type of matter to another (parts of the matter of elementary particles into energy), and vice versa. Apparently, the same effect, but on a smaller scale, should be observed in chemical transformations of substances of an exo- and endothermic nature, as well as biological ones, etc. However, these are much more difficult to notice due to smaller energy consumption upon the formation of bonds in these types of substance at a different level of material organization. In general, the law of conservation of mass and energy in substances [6] is demonstrated by figure 1.

According to figure 1, in general, the energy of the intra-structural interaction of the elements that make up various substances and matter of the microcosm (that is, the interaction or bonds of the elements or particles that make them up), attributed to the unit of their mass (specific energy), decreases as the latter increases. This can be represented as follows:

$$E / m = K, \tag{2}$$

where K is the proportionality coefficient, which takes into account the regular nature of the inverse dependence of the energy of intra-structural interaction of elements of respective level of matter on the mass of an object that forms this level.

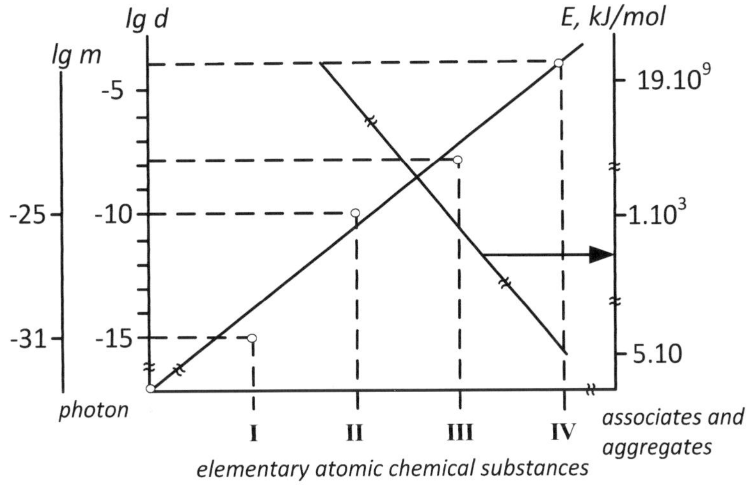

Figure 1. Size-mass and energy characteristics of the intrastructural interaction of particles of different levels of matter organization (units: mass *m* - kg, particle diameter *d* - m).

That is, *with an increase in the mass of various substances, upon transition (figure 1) from the lowest to the highest levels of substance organization (I – II – III – IV), the specific bond energy of its structural elements decreases. And, on the contrary, with an increase in the specific energy of the intra-structural interaction of substances, their mass decreases.*

As a result, "matter" today is no longer just a philosophical concept, but a natural science concept *characterized by mass and energy* [2, 3]. At the same time, matter exists in the form of two extreme basic varieties (forms) - *discrete* (discontinuous in the form of particles of substances and material bodies) and *continuous* (energy continuum). In accordance with the modern *paradigm of multilevel structural organization of matter*, transition from one of its variety to another occurs gradually when the ratio of size-mass and energy characteristics changes [1-3]. This happens due to material unity of nature of various objects of the universe and the presence of fundamental differences in their structure and properties within the framework of the *laws of conservation of mass – energy and particle-wave dualism.*

Thus, *energy* is a continuous (field) type of matter or *an energy continuum.* It exists in the form of open or closed (closed within discrete forms of matter) *fields*.

10^4 10^{-3} $8*10^{-7}$ $4*10^{-7}$ 10^{-8} 10^{-11} 10^{-13} m

RW	IR	Visible spectrum	UV	X-R	γ
Radio waves	Optical band			X-Rays	Gamma rays

Figure 2. Electromagnetic wave scale.

The most important distinguishing feature of the continuous forms of existence of the material energy continuum in the form of fields is possessing by the latter the wave characteristics, which vary in a fairly wide range. See, for example (figure 2), the scale of electromagnetic waves [5], while understanding that the wave characteristics of gravitational, nuclear, chemical and other fields would differ from the above electromagnetic ones.

As a result, the material energy continuum has a set of unique properties, among which two most important ones can be distinguished. The first property is the ability to determine the level of interaction between discrete forms of matter (substances and material bodies), up to the formation of stable bonds and the transformation of substances into larger and larger material formations. For example, elementary particles (substances) are transformed into substances with a higher mass-atoms (nuclear physical bonds), then atoms are bound into substances with an even higher mass – chemical substances (chemical bonds), then biological substances, material bodies, in the form of planets in star systems (gravitational bonds). The second property is the ability to determine the shape of the movement of material objects. For example, gravity, together with mass, determines the forms of motion of matter, primarily material bodies (planets, rockfalls, the direction of movement of rivers, etc.), that is, the mechanical forms of motion of matter. Further, it is necessary to distinguish the forms of motion of matter at the level of different substances. As a result, the *material energy continuum* has a set of unique properties, among which two most important ones can be distinguished. *The first property is the ability to determine the level of interaction between discrete forms of matter* (substances and material bodies), up to the formation of *stable bonds and the transformation of substances* into larger material formations. For example, elementary particles (substances) are transformed into substances with a higher mass – atoms (nuclear physical bonds), then atoms are bound into substances with an even higher mass – chemical substances (chemical bonds), then biological substances, material bodies in the form of planets in star systems (gravitational bonds). *The second property is the ability to determine the form of material objects' movement.* For example, gravity, together with mass, determines the forms of motion of matter, primarily material bodies (planets, rockfalls, the direction of movement of rivers, etc.), that is, the *mechanical forms of motion of matter*. Further, it is necessary to distinguish the forms of motion of matter at the level of different substances. The chemical bond (chemical energy continuum) in substances and its energy determines the features of the *chemical form of motion of substantial matter* in the form of chemical transformations. We would like to note the specifics of electric energy as a kind of chemical energy, since it is determined by the presence in the substance of a metallic type of chemical bond in the form of an "electron gas" - delocalized collectivized electrons in the internuclear space. Nuclear bonds in atoms (physical substances) and their energy determine the features of the physical form of motion of matter in the form of physical nuclear transformations. Thermal energy is an apparently mixed physicochemical type of motion of matter. This is due both to the movement of individual molecules relative to each other with the overcoming of intermolecular van der Waals or hydrogen bonds (for example, H_2O molecules in water), and to the change in the type of chemical bond upon heating due to the elements in the crystal lattice of a materials vibrating. A special case is the so-called kinetic energy (energy of motion), which, in fact, is no longer energy as such. It characterizes the wave properties in accordance with the de Broglie relation and the Einstein formula (1) for various forms of motion of matter and the magnitude of the energy characteristics depending on mass and speed of material objects' motion.

Since chemical energy is the main source of heat and electricity, the authors propose a new scientific foundation for the systematization and quantitative assessment of the energy characteristics of chemical compounds of elements (chemicals). It is based on a comprehensive assessment of the contribution of the composition of chemical compounds and their chemical bond type, based on a single model of chemical bond and the "System of chemical bonds and compounds" (SCBC) in the form of a "Chemical Triangle" (CT) [6-13].

The authors of the present paper placed in the triangle's vertices three extreme types of chemical bond (figure 3), while mixed homo- and heteronuclear bonds and compounds formed by these bonds are located on the left side (homonuclear bonds) and area (heteronuclear bonds) of this CT. Values of covalent character (C_C) for some homo- and heteronuclear bonds are given in brackets on the left and right sides of the CT (or, more precisely, on the axes parallel to the sides of CT). The left side of the CT represents homonuclear bonds of *s*-, *p*-, *d*- and *f*-elements (I), as well as respective basic metallic and nonmetallic (low-, oligo-, highmolecular compounds and polymer solids, like diamond) substances. Heteronuclear chemical bonds are characterized by three components and therefore occupy the area of the CT (e.g., Mg_3Sb_2, Fig. 3). However, for convenient presentation of various binary heteronuclear chemical compounds (like oxides, nitrides, carbides, etc.) within the CT they are located on additional dotted axes parallel to the right side of the CT (figure 3). Chemical bonds cannot be found at the bottom and right sides of the CT, as well as at M and I vertices (this is why these sides are shown as dotted lines) due to the fact that C_C cannot be equal 0.

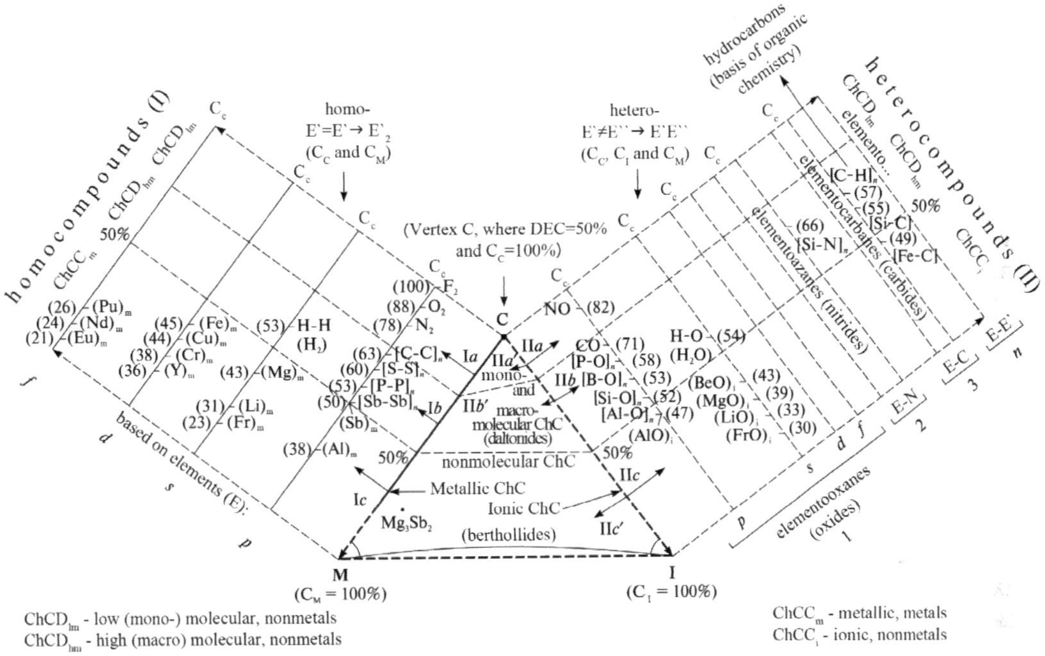

Figure 3. Chemical Triangle is a system of chemical bonds and compounds (SCBC) that combine homo- and heteronuclear binary chemical compounds in the form of nonmolecular metallic, ionic, mono-, oligo-, and macromolecular compounds, along with polymer bodies where ChCDs are discrete chemical compounds (molecular) and ChCCs are continuous chemical compounds (nonmolecular); C_C – covalent character of chemical bond, C_M – metallic character, C_I – ionic character.

In the same direction work is underway in the United States and Europe, but there are no attempts to quantify the metallic character, to develop a single model of chemical bonding [14-20] aiming at the creation of a truly chemical system in the form of a Chemical triangle [12, 13]. As a result, the use of this basic scientific innovation for the first time made a symbiosis of the approach of two outstanding Russian chemists: D. I. Mendeleev's periodic system of atoms (composition – property) and A. M. Butlerov's theory of chemical structure of matter (composition – chemical structure – property). As a result, this led the authors to create (Figure 3) a truly chemical system – the System of chemical bonds and compounds (SCBC) and a new paradigm for the systematic study of chemical substances (composition - chemical bond and structure - property) [6-13]. The foundation was laid for the creation of a database on the system digitalization and evaluation of the energy stored in various chemicals (gas, coal, fuel oil, peat, wood, etc.) and the most effective ways to extract it from them [21, 22]. The extreme relationship in terms of the influence of the composition and chemical bond type on the energy characteristics of substances depending on the gradual change in the ratio of the covalent and metallic bond components in homonuclear and additionally ionic bond component in heteronuclear compounds of elements was shown. The proposed approach and the SCBC can be used in power engineering, chemical industry, construction, industry of nanomaterials, etc. [7-23].

4. Conclusion

For the first time, it was shown that energy, as well as matter, objectively exists in the form of its various material varieties. Energy is a continuous (field) type of matter - the energy continuum. It exists in the form of open or closed (closed within discrete forms of matter) fields. They differ in structure, including the ability to perform different types of work, to determine the forms of interaction and movement of matter in various material systems (substances, material bodies and megamaterial systems). A new scientific foundation is proposed for the quantitative assessment, systematization and digitalization of the energy characteristics of the entire array of chemical compounds of elements (chemicals), based on the System of chemical bonds of compounds (SCBC) in the form of a "Chemical Triangle".

5. References

[1] Kuznetsov V I, Idlis G M and Gutina V N 1996 *Natural Science* (Moscow: Agar) (in Russian)

[2] Sirotkin O S 2011 *Integral-differential principles of unitary natural science (Paradigm of multi-level organization of matter as a natural basis for diversity and unity of nature of objects of the universe system)* (Kazan: KSPEU) (in Russian)

[3] Sirotkin O S 2020 *System of Universe (Scientific principles of the contemporary view of the world)* (Moscow: Ruscience) (in Russian)

[4] Feynman R, Leyton R and Sands M 2019 *Feynman physical lectures* (Moscow: AST) (in Russian)

[5] Karpenkov S H 2003 *Concepts of contemporary natural science* (Moscow: Vysshaya Shkola) (in Russian)

[6] Sirotkin O S, Sirotkin R O and Trubacheva A M 2005 *Russ. J. of Inorg. Chem.* **50** 67

[7] Sirotkin O S, Sirotkin R O and Trubacheva A M 2009 *Characteristics of homo- and heteronuclear bonds of fine electronic-nuclear structure and their impact on properties of metallic and nonmetallic materials* (Kazan: KSPEU) (in Russian).

[8] Sirotkin O S, Sirotkin R O and Shibaev P B 2011 *Russ. J. Inorg. Chem.* **56** 1104

[9] Sirotkin R O, Sirotkin O S, Ivshin I V, Safin A R, Tsvetkov A N and Dolomanyuk L V 2016 *ARPN J. of Eng. and Appl. Sci.* **11** 9664.

[10] Sirotkin R O and Sirotkin O S 2018 *IOP Conf. Series: Materials Science and Engineering* **327** 042119.

[11] Sirotkin R O and Sirotkin O S 2020 *Materials Science Forum* **992** 952.

[12] Sirotkin O S and Sirotkin R O 2019 *Chemistry (Principles of a Single Chemistry)* (KNORUS, Moscow) (in Russian).

[13] Sirotkin R O and Sirotkin O S 2020 *Russ. J. Phys. Chem. A* **94** 1153.

[14] Jensen W B 2009 *J. Chem. Educ.* **86** 278.

[15] Sproul G D 1994 *J. Phys. Chem.* **98** 6699.

[16] Sproul G D 1994 *J. Phys. Chem.* **98** 13221.

[17] Allen L C 1992 *J. Am. Chem. Soc.* **144** 1510.

[18] Jensen W B 1995 *J. Chem. Educ.* **72** 395.

[19] Leach M R 2013 *Found. Chem.* **15** 13.

[20] Kong F 1982 *J. Mol. Struct.* **90** 17.

[21] Sirotkin O S and Maloletkov P A 2014 *The system of homonuclear substances*. Certificate of state registration of computer programmes **20144617629** (in Russian).

[22] Sirotkin O S, Sirotkin R O, Pavlova A M, Buntin A E, Perukhin M. Yu, Khabibullina R R 2019 *Characteristics of chemical bonds and their influence on the structure and properties of simple substances and oxides*. Certificate of state registration of computer programmes **2019618435** (in Russian).

[23] Protopopov E A 2020 *Evaluation and forecasting of the working capacity of steels of oil and gas pipelines according to the results of the analysis of the specific parameters of interatomic bonds*. PhD thesis (Tula: Tula State University) (in Russian).

Study heat exchange in porous structures

A A Genbach[1], H I Beloev[2], D Yu Bondartsev[1*] and N A Genbach[1]

[1] Almaty University of Energy and Communications, Department of Heat & Power Units, 126/1Baitursynov Str., Almaty, Kazakhstan
[2] University of Ruse, Agricultural Machinery Department, 8 Studentska Str., Ruse, Bulgaria

*E-mail: d.bondartsev@aues.kz

Abstract. This paper shows the problem with heat exchange depending on units of thermal power plant equipment. The type of structures is determined and the heat flow for different pressures is proposed. Studies are developed for the condition of the heat exchange surface. Devices with porous coatings eliminate the development of cracks in the components and units of TPP equipment have been suggested. The research is applicable to gas turbine units of TPP. Comparable capillary-porous and flow systems have high reliability, but the former allowed the reduction of coolant consumption dozens up to 80 times. The results show that at higher heat loads it is suitable to use in porous surfaces to control the cooling surface. Evaluation of capillary-porous structures has shown their advantages over traditional cooling systems.

1. Introduction

The trends of industrial thermal power engineering in the oil industry, requires intensification of processes in new high-heat and high-pressure combustion chambers and nozzle apparatuses of TPP. The authors of contemporary works [1-18] are still interested in heat transfer of homogeneous and non-homogeneous porous coatings and from specially designed wicks to increase heat transfer. This study addresses issues of thermohydraulic characteristics of boiling in porous media - bubble size, departure frequency, and nucleation density, with the help of thermocouples and high-speed cameras. It has been found the most effective and reliable way to intensify heat exchange during boiling is the use of capillary-porous coatings. In this way, the operating mode and technological indicators are improved at a low price [19-25].

2. Experimental setup

Heat flows of the boiling process in porous structures were calculated using photo cinematography and holographic monitoring (figure 1) [19].

Figure 1. Experimental setup for studying the vaporization processes by optical methods: 1 - laser; 2 - cine-photocamera SKS-1M; 3 - cooling element with porous structure [22].

The elements from figure 2 are: the surface temperature, t_{st}, liquid consumption m^{t}_{l}, m_1, m_2, m_{dis}, m_c, $m_{c.w}$, m_{air}, steam m_{st}. Where: t - tank, dis – discharge, c – condensate, c.w. – condensing water, a – air. Temperatures of liquid, t_w, t_l^{t}, t_l^{dr}, t_l^{out}, t_l^{in}, steam t_{st}, electrical insulation $t_{el}^{i} = t_{dif}$.

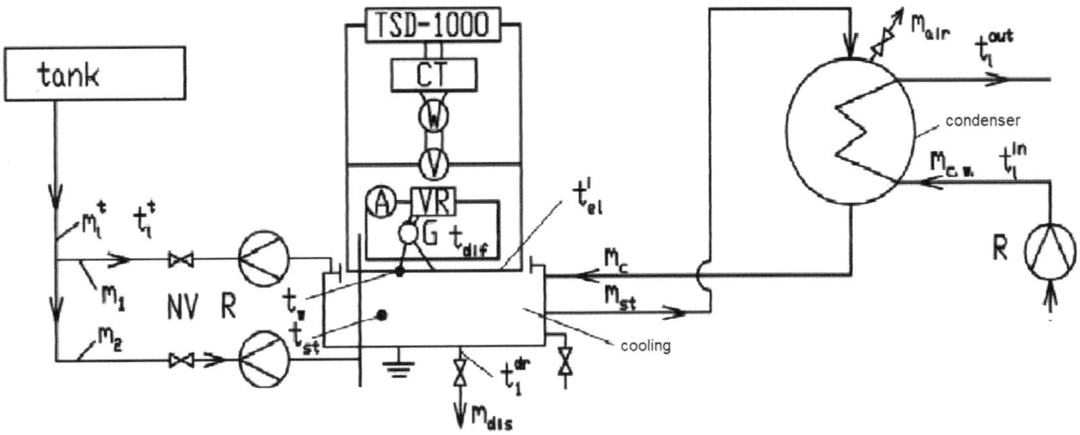

Figure 2. Principal scheme: TSD –transformer, CT – universal current transformer, W – power meter, V – voltmeter, A – ammeter, VR – voltage regulator, G – galvanometer, R – rotameter, NV – needle valve

The TSD welding transformer supplied energy to the heater. The current is measured with a current transformer type CT-6M2. The voltage of the heater is measured with a voltmeter type D523. The measurement accuracy is ± 1% for the current and ±1.6% for the voltage drop, respectively. With the VR protection heater, the current is maintained from 200 to 1200 A. TL-4 thermometers with a range of 0-50° C and 50-100°C are used respectively to measurement of the liquid and the environment temperature. Drainage fluid and vapor temperatures are measured using L-type thermocouples. It free ends are placed in melting ice. The thermocouple is connected to switches type PP-63. Consumption of coolant and/or circulating fluid is read by rotameters RED. Consumption of drained liquid and condensate is read with a pressure scale device 0.5×10^{-3} l and the filling time with a stopwatch type S-P-1b with a set 0.1 s/div [4].

The error in measuring the flow rate of liquid by rotameters must not exceed ± 3% but by volume method up to ± 2 [20]. The discrepancy between the material balance between the coolant flow rate and the drainage and condensate flow rate is up to ±10%. More information on measurement conditions and techniques is published in references [22-24].

3. Calculation of heat flows

As follows from photo-cinematic and holographic observations [20-23], the dynamic of the vapor phase after nucleation of a bubble of critical size proceeds with the participation of the evaporating micro-layer of liquid under the vapor bubble. Under certain conditions, followed by the forming of a "dry spot" [20]. The value of the detachment radius R_0 is calculated by the equation provided in [25]. After some "silence" of the generation, a new bubble spontaneously emerges, with growth time controlled by the coolant flow rate decreasing and occurs with a more intense heat supply from the thin superheated liquid layer. The incommensurable of the "silence" time with the growth period is also indicative of an existing superheated pulsating micro-layer of fluid, the durability, and stability of which is extended by the action of forces.

When the liquid boils in a big volume, the phase of bubble growth relate to forces ~ (10 up to $100)\times10^{-3}$ s, while in a porous structure this value is about ten times less. In both systems, the nucleation and detachment stages of the bubble are negligibly small and have explosion character. [24]. The silence time in the periodic cycle of bubble formation during large-volume boiling could be ~ 0.1 s and be commensurate with the bubble growth time.

As a result occurs the exposure of a nucleus of critical size. The increasing bubble surface slows down its expansion in the depression. Heat is transferred mostly through the micro-layer of liquid under the vapor bubble, which is cone-shaped with a "dry" spot of the bubble. The main evaporation process occurs at the base of the "dry" spot. The micro-layer of the "dry" spot change insignificantly during bubble growth [19] due to the inflow of fresh portions of coolant transported by gravitational and capillary forces, due to the lifting forces acting on the bubble. The inertial forces are hapened in the initial time of bubble grow and decrease before bursting. The bubble acquires a spherical shape of the "dry" spot and it's reduced, and the shape of the micro-layer changes significantly. Bubble detachment is depends from surface and hydro-gasodynamic resistance forces, and the excess of liquid. Thus through the underheating and flow rate of the liquid is created and the control of thermohydraulic characteristics is realized.

When the bubble upper boundary touches the porous structure surface, the bubble collapses. According to [19], a hole appears at the point of contact, through which steam flows out of the bubble into the vapor volume. A liquid wave starts to spread along the surface. The process described proceeds (10^{-7} up to 10^{-6}) s, i.e., it has an explosive wave character, character, the same as the "birth" of a vapor nucleus, with the outflow of the light phase. This phenomenon has been used to draw an analogy with explosive processes in elliptical systems [23], to separate energy processes for wave energy and compressed gases. The phenomenon described was accompanied by intense ejection of liquid droplets [22], but no disruption of the porous system was observed.

A relatively cold liquid rushes into the depression and some of the vapor there condenses until the vapor and liquid temperature at the limit of their contact is equalized. When the liquid excess is large, all the vapor can condense and the bubble-forming effect will stop. The inflow of accumulated heat in the wall will lead to a new generation cycle of nucleation. During one cycle the wall temperature below it will change which shows the high heat transfer intensity. Reducing the heat flow will return the system to its previous state.

The bubble shape growing is a form between spherical and hemispherical. The spherical bubble takes place when water boils at high pressures (P>1 bar), and the hemispherical one - at low pressures (P<1 bar). At the initial moment the surface tension forces take over and the bubble transform into cells in the ball shape.

The bubble rapid growth at the initial moment is set by a fixed value of q. The described model allows describing experimental the mechanism and to obtain calculated dependences for liquid boiling of a porous structure [19]:

a) surface boiling area:

$$q = 1.1 \cdot 10^{-3} \cdot \lambda_{ef}\left(\alpha J_a {R_0}^{-2} v^{-1}\right)^{0.5} \cdot \Delta T F K^{-1} \quad (1)$$

$$F = (1 + cos\beta)^{0.57} \cdot m^{0.25} \cdot \left(\frac{h}{\delta}\right)^{0.5} \cdot \left(\frac{b_0}{b_r}\right)^{0.27},$$

where $\boldsymbol{J_a} = \frac{c\rho_l(T_w - T_p)}{r\rho_v}$ is Jacob's number and $\boldsymbol{K} = \boldsymbol{1} + \left[\frac{(\rho c\lambda)_l}{(\rho c\lambda)_w}\right]^{\boldsymbol{0.5}}$ – wall coefficient.

This equation is useful for calculating cooling systems for steam and gas turbines in TPP [19, 23].

b) area of bubble boiling developed:

$$q = 0.0862 \cdot r \cdot \rho_v \cdot \alpha \cdot J_a \sqrt{n \cdot F \cdot K^{-1}} \quad (2)$$

c) evaporation crisis ($\boldsymbol{b_r} > \boldsymbol{0.28 \cdot 10^{-3}}$ m):

$$q_{cr} = 0.0347 \cdot r[g \cdot (\rho_l - \rho_v) \cdot \rho_v \cdot D_0]^{0.5} \cdot \left(\frac{b_r}{b_0}\right)^{0.3} \cdot \left(\frac{\delta}{\delta_0}\right)^{0.5} \cdot (1 + cos\beta)^{0.6} \quad (3)$$

The analysis shows that the steam bubble growth on the wall leads to an increase in temperature and decrease in efficiency.

In [19-25] are summarized the experimental data on different thermohydraulic processes. Calculated of heat load (equation 1-3) on temperature pressure, liquid excess, pressure, thermophysical properties of the wall, orientation, geometric characteristics of porous structures and heat exchange wall are obtained. These equations allow for calculations in the design of TPP.

4. Data analysis

Two types of evaluation criteria have been approved - dimensionless number St and dimensionless temperature. The obtained three areas were distinguished: the area of high values of the dimensionless temperature - from 1.3 to 1.7; the area of moderate values - from 1.0 to 1.3 and the area of low values - from 1.0 and less. The dimensionless temperature has been adopted as a criterion for this division.

Area of low values. Decreasing the temperature to such values is unacceptable and when the temperature reaches the value of 1.0 a boiling crisis occurs.

Area of moderate values. The operating range for capillary-porous materials (c.p.m.) from 1.0 to 1.3. At small values, the priority of single-layer c.p.m. becomes obvious, from 1.0 to 1.2. This type of heat exchange process is characteristic of combustion chambers. If the range 1.2-1.25 values is reached, the priority of one or another c.p.m. should be determined. For this purpose, an additional criterion St is used. In the range 1.25-1.3 the priority of another c.p.m. should be found on the base on the distribution of the intensity of the heat transfer process over the evaporator surface and possible wall temperature fluctuations. The area of the dimensionless temperature is divided into three-zone: from 0.1 to 1.0×10^{-3} m; from 1.0 to 2.0×10^{-3} m and from 2.0×10^{-3} m.

5. Conclusions

The correlation between the heat transfer coefficient and the wall temperature is inversely proportional. For a traditional (flow-through) cooling system, the allowable temperature fluctuations are no more than 10 K (2.5-3%), with all heat flow accumulated by the wall, leads to temperature overstress in the material. Any increase in temperature requires an increase in cooling. For a porous cooling system, the maximum allowable temperature fluctuations, the cooling system compensates for independently, and with a high degree of response to changes in the specific heat flow, are up to 138.9 K, and the loss of heat flow will be 1.174 MW/m^2. Considering that the working temperature is 190.15 ^{0}C for heat flow of 13 MW/m^2 the system's temperature reserve is 73%, and the heat flow reserve is 11%.

6. References

[1] Boubaker R and Platel V 2016 Dynamic model of capillary pumped loop with unsaturated porous wick for terrestrial application. *Energy*. **111**, pp 402-413 https://doi.org/10.1016/j.energy.2016.05.102

[2] Lei G, Li W and Wen Q 2019 The convective heat transfer of fractal porous media under stress condition. *International Journal of Thermal Sciences*. **137**, pp 55-63 https://doi.org/10.1016/j.ijthermalsci.2018.11.017

[3] Mori S and Okuyama K 2009 Enhancement of the critical heat flux in saturated pool boiling using honeycomb porous media. *International Journal of Multiphase Flow*. **35**, pp 946-951 https://doi.org/10.1016/j.ijmultiphaseflow.2009.05.003

[4] Poniewski M E 2004 Peculiarities of boiling heat transfer on capillary-porous coverings. *International Journal of Thermal Sciences*. **43** 5, pp 431-442 https://doi.org/10.1016/j.ijthermalsci.2003.10.002

[5] Kimihide O and Nagano H 2019 Investigation on liquid-vapor interface behavior in capillary evaporator for high heat flux loop heat pipe. *International Journal of Thermal Sciences*. **140**, pp 530-538 https://doi.org/10.1016/j.ijthermalsci.2019.03.008

[6] Ji X et at 2013 Pool boiling heat transfer on uniform and non-uniform porous coating surfaces. *Experimental Thermal and Fluid Science*. **48**, pp 198-212 https://doi.org/10.1016/j.expthermflusci.2013.03.002

[7] Chang Y H and Ferng Y M 2019 Experimental investigation on bubble dynamics and boiling heat transfer for saturated pool boiling and comparison data with previous works. *Applied Thermal Engineering*. **154**, pp 284-293 https://doi.org/10.1016/j.applthermaleng.2019.03.092

[8] Chuang T J, Chang Y H and Ferng Y M 2019 Investigating effects of heating orientations on nucleate boiling heat transfer, bubble dynamics, and wall heat flux partition boiling model for pool boiling. *Applied Thermal Engineering*. **163**, 114358 https://doi.org/10.1016/j.applthermaleng.2019.114358

[9] Khan S, Sezer N and Koç M 2019 Design, fabrication and nucleate pool-boiling heat transfer performance of hybrid micro-nano scale 2-D modulated porous surfaces. *Applied Thermal Engineering*, **153** 5 pp 168-180 https://doi.org/10.1016/j.applthermaleng.2019.02.133

[10] Hanlon M A and Ma H B 2003 Evaporation heat transfer in sintered porous media. *Journal of Heat Transfer* **125** 4 pp 644-652 https://doi.org/10.1115/1.1560145

[11] Arik M, Bar-Cohen A and You S M 2007 Enhancement of pool boiling critical heat flux in dielectric liquids by microporous coatings. *International Journal of Heat and Mass Transfer* **50** pp 997-1009 https://doi.org/10.1016/j.ijheatmasstransfer.2006.08.005

[12] Sarwar M S, Jeong Y H and Chang S H 2007 Subcooled flow boiling CHF enhancement with porous surface coatings. *International Journal of Heat and Mass Transfer* **50** pp 3649-3657 https://doi.org/10.1016/j.ijheatmasstransfer.2006.09.011

[13] Alekseik O S, Kravets V Yu 2013 Phisycal model of boiling process on porous surface in limited space. *Eastern-European Journal of Enterprise Technologies*, **8** (64) pp 26–31 https://doi.org/10.15587/1729-4061.2013.16448

[14] Li J, Hong F, Xie R and Cheng P 2019 Pore scale simulation of evaporation in a porous wick of a loop heat pipe flat evaporator using Lattice Boltzmann method. *International Communications in Heat and Mass Transfer* **102** pp 22-33 https://doi.org/10.1016/j.icheatmasstransfer.2019.01.008

[15] Jamialahmadi M et at. 2008 Experimental and theoretical studies on subcooled flow boiling of pure liquids and multicomponent mixtures. *International Journal of Heat and Mass Transfer* **51** pp 2482-2493 https://doi.org/10.1016/j.ijheatmasstransfer.2007.07.052

[16] Ose Y and Kunugi T 2011 Numerical Study on Subcooled Pool Boiling *ASME/JSME 2011 8th Thermal Engineering Joint Conference* (Honolulu, Hawaii, USA) **T10193** https://doi.org/10.1115/AJTEC2011-44401

[17] Kuzma-Kichta Y A and Leontiev A I 2018 Preface: heat and mass transfer enhancement on macro-, micro-, and nanoscales *Journal of Enhanced Heat Transfer* **25** doi: 10.1615/jenhheattransf.v25.i6.10

[18] Choi E Y et at. 2013 Heat transfer augmentation using a rib–dimple compound cooling technique *Applied Thermal Engineering* **51** pp 435-441 https://doi.org/10.1016/j.appltherm aleng.2012.09.041

[19] Genbach A, Bondartsev D, Iliev I and Georgiev A 2020 Scientific method of creation of ecologically clean capillary-porous systems of cooling of power equipment elements of power plants on the example of gas turbines. *Energy* **199**, 117458. https://doi.org/10.1016/j.energy.2020.117458

[20] Genbach A, Bondartsev D and Iliev I 2019 Heat transfer crisis in the capillary-porous cooling system of elements of heat and power installations *Thermal Science* **23** pp 849–860 https://doi.org/10.2298/TSCI171016139G

[21] Genbach A, Bondartsev D and Beloev H 2021 Studies of integrated natural capillary-porous coatings, 2021, *IOP Conf. Ser.: Mater. Sci. Eng.* **1032** 012038 https://doi.org/10.1088/1757-899X/1032/1/012038

[22] Genbach A, Bondartsev D and Beloev H 2021 A study of thermal devices with porous coatings, 2021 *IOP Conf. Ser.: Mater. Sci. Eng.* **1032** 012039 https://doi.org/10.1088/1757-899X/1032/1/012039

[23] Genbach A and Bondartsev D 2020 Limiting thermal state of capillary-porous power-plant components. *Russian Engineering Research.* **40** pp 384-389 https://doi.org/10.3103/S1068798X20050093

[24] Genbach A, Shokolakov K, Bondartsev D and Iliev I 2019 Research of the heat transfer crisis and limit state of the heat exchange surface covered by capillary porous medium, 2019, *IOP Conf. Ser.: Mater. Sci. Eng.* **595** 012007 https://doi.org/10.1088/1757-899X/595/1/012007

[25] Genbach A, Bondartsev D, Iliev I and Terziev A 2019 Analogy in the processes of heat exchange of capillary-porous coatings in energy installations. *E3S Web Conf.* **85** 05003 doi: https://doi.org/10.1051/e3sconf/20198505003

Heat transfer for the boiling crisis in porous systems

A A Genbach[1], H I Beloev[2], D Yu Bondartsev[1] and N A Genbach[1]
[1] Almaty University of Energy and Communications, Department of Heat & Power Units, 126/1 Baitursynov Str., Almaty, Kazakhstan
[2] University of Ruse, Agricultural Machinery Department, 8 Studentska Str., Ruse, Bulgaria

E-mail: d.bondartsev@aues.kz

Abstract. In this paper we analized and investigated the heat exchange crisis of boiling in porous structures, applicable in thermal power plants. Then we describe the heat exchange processes mechanism and determined the ideal sizes and thicknesses of porous structures. The designed porous structures can be implemented in gas turbine's nozzles and combustion chambers. From an environmental point of view, the consumption coolant liquid is reduced by ten times in comparison the standard flow system. It's effectively to develop mesh structures to allow the extension of the critical loads and manage the surface border.

1. Introduction

The advanced technologies play a significant part in the mechanical engineering – nanotech and plasma processing allows of creation the porous coatings on the product's surfaces. They can serve as a tank and a micro-channels for lubricants, ensuring operational efficiency of rubbing body parts [1-8].

One of the promising and effective methods of intensification of heat exchange is the use of highly conductive capillary-porous and mesh materials (PSM). Such materials have long been used in liquid-propellant rocket engine (LPRE), as filters, capillary intake devices, in the manufacture of permeable firing bottoms of mixing heads and other devices [9-10].

In the production of porous materials one of the main characteristics of processing and application is stated by the structure of the pores of the material. The porous materials are robust and resistant to corrosion, with wide operating temperature range, don't clog the filtered liquid, machinability, allow multiple regeneration, have high electrical and thermal conductivity [11-15].

2. Experimental setup

In our research we used a rocket-type flame-jet burner (marked with 1 on figure 1) with a porous cooling system 2 and with 3 is denoted the casing. With such rocket – type jet burner can be produced different heat transfer intensifiers for cooling of nozzles and combustion chambers [13, 15].

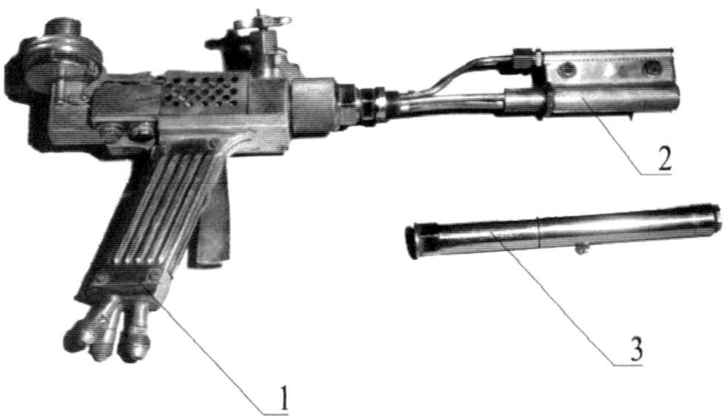

Figure 1. Rocket-type flame-jet burner.

Burner (figure 2) Combustion chamber 1, a Laval nozzle 2 with confusor 3 and diffuser 4 parts and critical section 5 between them, swirl 6, atomizer 7 and distribution head 11 with branch pipes 12, 13, 14 for supply of oxidizer 19 and fuel 20 to combustion chamber 1 and coolant 18 to the cooling cavity 10 formed between shell 8 and the outer walls of combustion chamber 1 and Laval nozzle 2, installed in the housing 8. The casing 8 has holes 9, 21 for outlet of the coolant 18. The diffuser part 4 of the Laval nozzle 2 is provided with an insert 15 having a capillary-porous coating 16 on the side facing the cooling cavity 10.

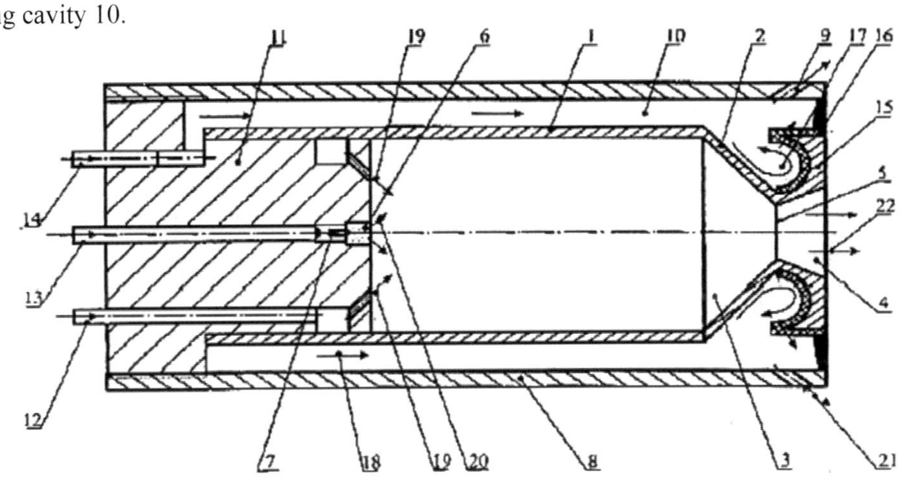

Figure 2. Burner cooling system for high pressure detonation combustion mode.

Number 16 (capillary-porous coating) is produced by a multilayer grid, the layers cells size increases with 0.08; 0.14 and 1.0 mm in the way of the outer surface of the coating. The radius of curvature of groove 17 is 2.5-5 times greater than its depth. Combustion products 22 are drained from the diffuser part 4 of nozzle 2.

The coolant lubricants 18 is passed through the branch pipe 14, in this way the coolant lubricants moving toward the cavity cooling 10. The coolant lubricants cools the combustion chamber 1 and the nozzle 2, then is discharged outside via the holes 9 and 21. Oxidizer 19 is mixed with the fuel 20, ignited by an ignition source and the resulting mixture is burned.

In the combustion chamber 1 a flow is formed with a temperature above (2000÷2500) ^{0}C and a speed of about 1600÷2000 m/s (in Laval nozzle 2) thus processing combustion products 22 [9].

Making the radius of curvature of groove 17 2.5÷5 times bigger than its depth redound to the elimination of thermal stress concentration and indemnification of thermal distention in the thermal groove. High capillary pressure is created if the capillary-porous coating 16 is isotropic, i.e. the size of the cell increases towards the outer surface. [14]

The listed advantages are realized both at nominal load and at alternating and emergency operation modes (no blockage of the cooling cavity by steam bubbles). Coolant liquid consumption is significantly reduced four times.

3. Experiment conditions

The main heater was made of nichrome foil, with thickness 0.05; 0.1; 0.3; 0.5; $0.7 \cdot 10^{-3}$ m, or stainless-steel plate – 1 and $2 \cdot 10^{-3}$ m. The heater size was from 0.05 m to 0.3 m with height 0.15 m to 0.7 m. The different structures were prepared of woven brass, copper and stainless steel woven meshes, composed of several layers (max 3). For the cooling surface we used 12X18H9T or 18H1OT stainless steel type, copper M2, brass L8O, bronze, alundum, nickel and glass. The minimum thickness of the wall was 0.05 and the maximum $2 \cdot 10^{-3}$ m. Before the experiments, the cooled wall and structure were cleaned and degreased.

The SKS-1M high speed camera was used for observations of the evaporation processes. Experimental installations, conditions and experimental data processing methods are described in [9-12, 14].

In the experiments we used both water with a pressure of 0.001 to 1 MPa and an aqueous solution of the foaming agent type PO-1.

The fluid velocity varies from $1.1 \cdot 10^{-3}$ to 0.1 m/s, underheating 0 to 20 K, fluid excess was from 1 to 14 from the overall steam consumption. The heat load $1 \div 60 \cdot 10^4$ W/m^2, thermal pressure $1 \div 60$ K, orientation of the system was relative to the verticals $\pm 0^0 \div \pm 90^0$.

4. Data analysis

For the experimental data analysis we used transitional boiling mode with heat transfer flows q = $(1 \div 8) \cdot 10^4$ W/m^2. We found a significant effect of coolant-flow rate and type of structure for this type of boiling mode. This applies to grids with a large mesh size. The incensement of the value q is associated with a decreasement of the amount of the heat-transfer coefficient and the appearance of vapor bubbles and also with an incensement of the thermal resistance of the boundary layer.

No restriction in the heat capacity rate of the system for different sizes of the cells. Hence in the case of heat pipes, the increased hydraulic resistance of the structure or the decrease in capillary absorption does not restrict the fluid flow.

A one-layer 0.14 structure or a two-layer 0.14 structure may be an acceptable option 0.4 and 0.55. Large cell sizes reduce fluid cleaning requirements and hydrogasdynamic resistance.

Increasing the coolant flow leads to redistribution of the flushing and boiling streams. For this reason, it is necessary to create a uniform temperature field along the height and length of the heat exchange surface through the concept of optimal excess liquid.

The ideal flow rate is defined as follows: up to $q \leq 100$ kW/m^2, the fluid flow rate is kept close to the heat pipes operation mode at the hydraulic diameter $0.28 \cdot 10^{-3}$ m and 30%, for values $> 0.28 \cdot 10^{-3}$ m; at $q > 100$ kW/m^2 exceeding the flow rate $1.3 \div 2$ times for values $\leq 0.28 \cdot 10^{-3}$ m allows to increase the range of removal of the value of q in $2 \div 3$ times in contrast with heat pipes.

On another hand the heat flow intensity is decreased with 40% for $q > 100$ kW/m^2, or the heat pipes are non-operational. The high intensity of the heat pipes can be obtained at $q < 20$ kW/m^2.

5. Conclusions

From the conducted research it's evident that with the used of capillary-porous material and value of the dimensionless temperature in the range from $1.0 \div 1.3$ the cooling system can be in normal operation mode. For the proposed burner device the optimum working area is accepted to be in the range of dimensionless temperature $1.25 \div 1.30$ (working range). In the operating range $\Delta q = 8.5 - 10.2$ MW/m^2 and operating interval $\Theta = 1.25 - 1.30$, the following capillary-porous materials are possible: combined with a ratio of width of cells of a parietal layer and the subsequent 0.7…($0.7 \times 1.0 \cdot 10^{-3}$ m);

single-layer, with cell width $1.0 \cdot 10^{-3}$ m; two-layer (with a layer thickness in width of a wire) with width of a cell $0.02 \cdot 10^{-3}$ m. From the technical point of view a two-layer capillary-porous structure with a cell width $0.02 \cdot 10^{-3}$ m is hard to achieve. To define the best capillary-porous material with the determination of the ideal combination of the cooling system, an analysis by the dimensionless criterion St is required. Determination of the characteristics and the kind of capillary-porous material must be done, in the general case, by at least two criteria: dimensionless temperature and St.

6. References

[1] Khan S A, et al. 2019 Design, fabrication and nucleate pool-boiling heat transfer performance of hybrid micro-nano scale 2-D modulated porous surfaces. *Applied Thermal Engineering.* **153** pp 168-180 https://doi.org/10.1016/j.applthermaleng.2019.02.133

[2] Wang W, et at 2013 Cooling performance analysis of steam cooled gas turbine nozzle guide vane. *International Journal of Heat and Mass Transfer.* **62** pp.668-679 https://doi.org/10.1016/j.ijheatmasstransfer.2013.02.080

[3] Wang W 2015 Efficiency study of a gas turbine guide vane with a newly designed combined cooling structure. *International Journal of Heat and Mass Transfer.* **80** pp 217-226 https://doi.org/10.1016/j.ijheatmasstransfer.2014.09.024

[4] Yang X, et at 2019 Turbine platform phantom cooling from air foil film coolant, with purge flow. *International Journal of Heat and Mass Transfer.* **140** pp 25-40 https://doi.org/10.1016/j.ijheatmasstransfer.2019.05.109

[5] Moon S W et at 2018 A novel coolant cooling method for enhancing the performance of the gas turbine combined cycle. *Energy.* **160** pp 625-634 https://doi.org/10.1016/j.energy.2018.07.035

[6] Ji X et at. 2013 Pool boiling heat transfer on uniform and non-uniform porous coating surfaces. *Experimental Thermal and Fluid Science.* **48** pp 198-212 https://doi.org/10.1016/j.expthermflusci.2013.03.002

[7] Chang Y H and Ferng Y M 2019 Experimental investigation on bubble dynamics and boiling heat transfer for saturated pool boiling and comparison data with previous works. *Applied Thermal Engineering.* **154** pp 284-293 https://doi.org/10.1016/j.applthermaleng.2019.03.092

[8] Chuang T J, Chang Y H and Ferng Y M 2019 Investigating effects of heating orientations on nucleate boiling heat transfer, bubble dynamics, and wall heat flux partition boiling model for pool boiling. *Applied Thermal Engineering.* **163** 114358 https://doi.org/10.1016/j.applthermaleng.2019.114358

[9] Genbach A, Bondartsev D, Iliev I and Georgiev A 2020 Scientific method of creation of ecologically clean capillary-porous systems of cooling of power equipment elements of power plants on the example of gas turbines. *Energy.* **199** 117458 https://doi.org/10.1016/j.energy.2020.117458

[10] Genbach A, Bondartsev D and Iliev I 2019 Heat transfer crisis in the capillary-porous cooling system of elements of heat and power installations *Thermal Science.* **23** pp 849–860 https://doi.org/10.2298/TSCI171016139G

[11] Genbach A, Bondartsev D and Beloev H 2021 Studies of integrated natural capillary-porous coatings, 2021, *IOP Conf. Ser.: Mater. Sci. Eng.* **1032** 012038 https://doi.org/10.1088/1757-899X/1032/1/012038

[12] Genbach A, Bondartsev D and Beloev H 2021 A study of thermal devices with porous coatings, 2021 *IOP Conf. Ser.: Mater. Sci. Eng.* **1032** 012039 https://doi.org/10.1088/1757-899X/1032/1/012039

[13] Genbach A and Bondartsev D 2020 Limiting thermal state of capillary-porous power-plant components. *Russian Engineering Research.* **40** pp 384-389 https://doi.org/10.3103/S1068798X20050093

[14] Genbach A, Shokolakov K, Bondartsev D and Iliev I 2019 Research of the heat transfer crisis and limit state of the heat exchange surface covered by capillary porous medium, 2019, *IOP Conf. Ser.: Mater. Sci. Eng.* **595** 012007. https://doi.org/10.1088/1757-899X/595/1/012007

[15] Genbach A, Bondartsev D, Iliev I and Terziev A 2019 Analogy in the processes of heat exchange of capillary-porous coatings in energy installations. *E3S Web Conf.* **85** 05003 DOI: https://doi.org/10.1051/e3sconf/20198505003

IOP Conf. Series: Materials Science and Engineering 1216 (2022) 012013 doi:10.1088/1757-899X/1216/1/012013

Optimization techniques for p-GTO thyristor design

M Cristea[1], F Babarada[1]

[1] Electronics Department, University Politehnica of Bucharest, 313 Spl. Independentei, Romania

E-mail: miron.cristea@upb.ro

Abstract. A new type of semiconductor power device was devised in the early '90s as an alternative to the classic Gate Turn-Off (GTO) thyristor. Because the low-doped n-base was replaced by a low-doped p-base, it was called the p-GTO. Its main advantage is a higher possible control voltage when the device is switched off, leading to the possibility of a higher blocking anode current (IATO) and a lower turn-off time. The studies and techniques employed with the help of SILVACO-TCAD simulation software Athena and Atlas show that the p-GTO has higher breakdown voltages compared with its classic counterpart and similar on-state voltage (VT) and switching characteristics when replacing the GTO in the same circuit. Specific circuit improvements, like an affordable higher turn-off gate voltage, will drive the p-GTO into even faster switching operation.

1. Introduction

The GTO thyristor is a p-n-p-n structure which can be switched-on and blocked with positive and respective negative signals applied on the gate, without reversing the voltage applied between anode and cathode [1,2]. The GTO thyristor is usually used as power electronic switch controlled by signals applied on the gate [3]. To improve the parameters of such devices by design means optimising toward the ideal electronic switch which has: null on-resistance, infinite off-resistance, sensibility for the control signal applied on the gate, noise immunity, compact device, high operating frequency (low switching times on-off), high reliability. The most important GTO thyristor performances are the maximum voltage and current and turn-off dynamic characteristics. They are function of device vertical structure parameters like thickness and doping of the semiconductor layers, surface geometry and thermal parameters. The geometrical configuration of gate-cathode architecture is very important to ensure a high rate of gate current evacuation in the turn-off process. The lifetime of the mobile electric charge carriers has a very important role to reduce the blocking time and affect the main device parameters. The thermal parameters of the case influence the thyristor switched power and the semiconductor chip temperature variations influence the device reliability.

2. Fabrication process

The main process (figure 1) starts with the anode (A) realised on boron doped with 10^{18} cm^{-3} concentration silicon wafer. Epitaxially growth of n-base of 2 microns silicon layer doped with phosphorous 10^{16} cm^{-3} concentration is followed by p-base 23 microns epitaxially growth of silicon layer doped with boron 10^{14} cm^{-3} concentration, connected to gate (G). The cathode (K) is processed by 0.2 microns epitaxially growth of silicon layer high doped with boron 10^{19} cm^{-3} concentration. The cathode-gate junction is configured by lithographic process and anisotropic etching of 0.2 microns windows in the 0.2 microns-thick epitaxially growth silicon layer highly doped with boron of 10^{19} cm^{-3} concentration. Same thickness silicon dioxide is deposited by chemical vapour deposition technique and lithographic patterned and by etching configured to fill the hole resulting by silicon anisotropic etching. The next process step is the 0.1 microns aluminium layer deposition, lithographic patterned and etching configuration of cathode and gate electrodes.

3. The gate-cathode (GK) junction optimization

Because the p-GTO has a low doped p-n junction at the cathode side (unlike the classic GTO) [4], with the help of Athena and Atlas TCAD and simulation software tools of SILVACO [5,6], the structure configuration was optimized on the GK-junction in order to minimize the electric field value at the surface of the junction and in the overall structure, in order to maximize the maximum applicable voltage on the Gate (reverse voltage) and on the Anode (direct voltage) with respect to the Cathode (usually as reference voltage of zero value). First step of process simulator begins with mesh definition [7]:

```
LINE X LOC=0.00 SPAC=0.1
LINE X LOC=0.4 SPAC=0.01
LINE X LOC=0.6 SPAC=0.01
LINE X LOC=1.00 SPAC=0.05
LINE Y LOC=0.0 SPAC=0.02
LINE Y LOC=0.2 SPAC=0.05
```

to refine the high doping gradient junctions, figure 1:

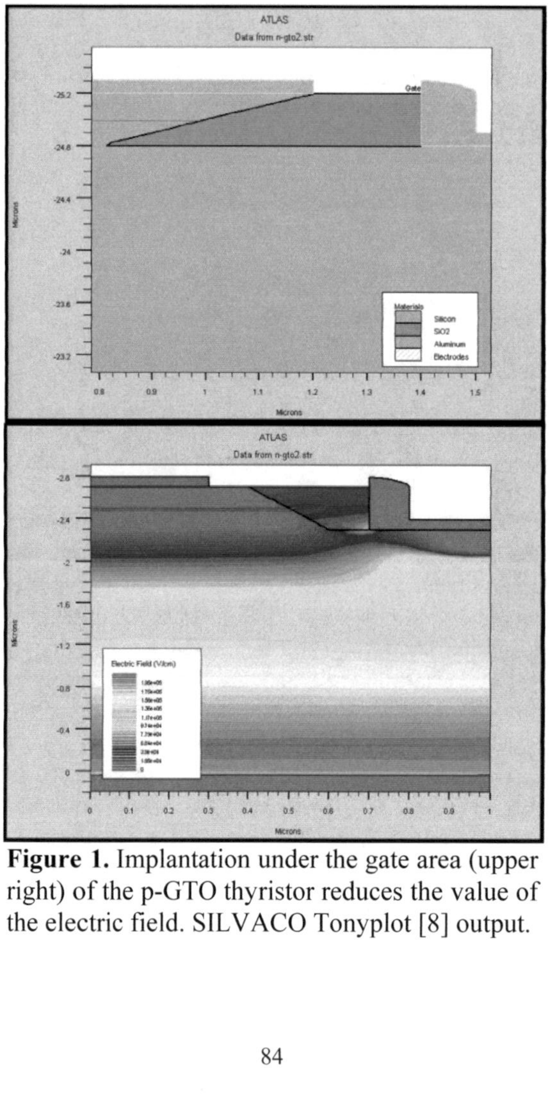

Figure 1. Implantation under the gate area (upper right) of the p-GTO thyristor reduces the value of the electric field. SILVACO Tonyplot [8] output.

IOP Conf. Series: Materials Science and Engineering 1216 (2022) 012013 doi:10.1088/1757-899X/1216/1/012013

For the sake of comparison, it was considered a classic GTO thyristor, with a p+n-pn+ structure and a p-GTO with p+np-n+ structure in which the concentration of the lightly-doped n-, respectively p- bases was the same (10^{13} cm-3 as in SILVACO simulation examples) – figure 2.

Two methods were employed to minimize the electric field and so to maximize the breakdown voltage at the GK junction:

- p-doping implant (Boron dose 4×10^{12} cm^{-2}) was added under the gate contact area, lowering the value of the electric field in that area - figure 2,

- positive bevel of an optimum angle of 45° at the gate-cathode junction (figure 1), decreasing the value of the electric field at the junction surface. Figure 1 shows a negative bevel at the GK junction since the upper side of the junction respectively the cathode is higher doped than the gate. The technological feasibility of the bevel is needed to achieve this optimisation, usually achieved by special technical processes.

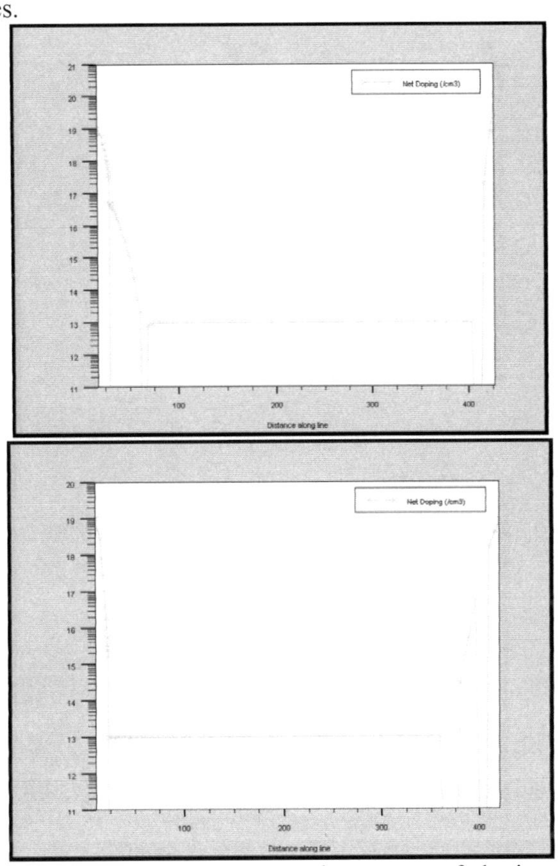

Figure 2. SILVACO Tonyplot output of doping profiles regions from right to left a) GTO thyristor: p+n-pn+ and b) p-GTO thyristor: p+np-n+.

4. Comparative results regarding the maximum voltage at the gate-cathode junction

Whereas the maximum (breakdown) voltage of the gate-cathode junction of the comparative classic GTO was obtained 9 volts, the corresponding voltage in the case of the p-GTO was 14.5 volts, which represents an increase of more than 50% in value. Also, in the figures can be observed that the rising of the leakage current is slower in the case of the p-GTO, which usually indicates a better designed junction.

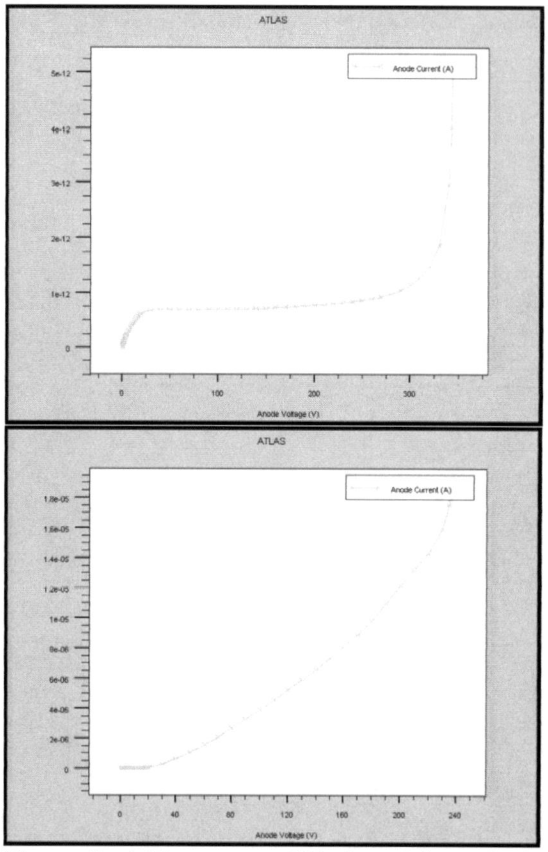

Figure 3. a) The reverse I/V plot of the p-GTO thyristor in comparison with b) the same for a classic GTO thyristor. SILVACO Tonyplot output.

After gate-cathode junction optimization (described above) of the p-GTO structure, the obtained value of the breakdown voltage of the gate-cathode junction was 27 V in the case a positive 45° bevel angle, compared with about 14 V in the cases of negative 45° or vertical 90° angles.

5. Comparative results regarding the anode to cathode maximum voltage

Also, in the case of the V_{AK} maximum voltage, a better result was obtained in the case of the p-GTO compared to the classic GTO: 340 vs. 240 volts, figure 3 a) and b). Since the doping of the high-resistive base of the devices was the same, this gain is due entirely to the novel structure of the p-GTO.

6. Comparative results about the switching characteristics

To properly compare the switching behaviour of the p-GTO versus the classic GTO thyristor, in both cases was used the very same circuit and simulation procedure provided in the power examples of SILVACO TCAD software. The only modification was to increase the anode inductance from 2.2 μH to 4 μH, because the old value was giving some minor oscillations on the descending front of the anode voltage, increasing too much the computation time of the simulation.

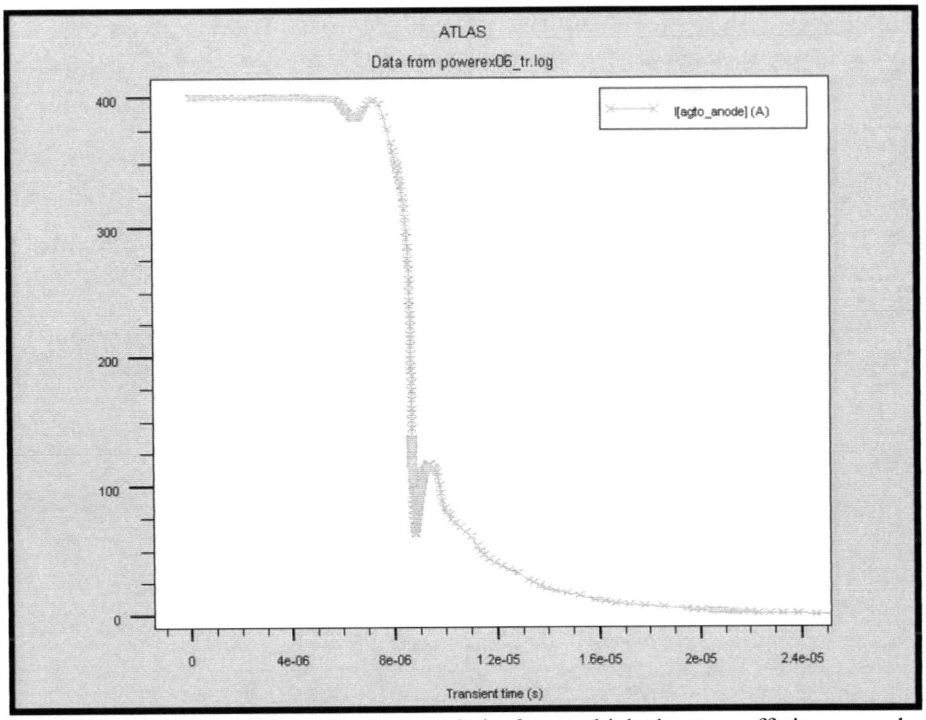

Figure 4. The turn-off current characteristic from which the turn-off times can be measured.

The initial value of the current to be switched off is 400 Amps. The turn-off time is similar in both cases: 6.4 μs for the GTO (ts = 2.3 μs + tf = 4.1 μs) and 6.7 μs for the p-GTO (ts = 2.3 μs + tf = 4.4 μs). Considering supplementary that in fact for the p-GTO a larger turn-off control voltage can be applied on the gate terminal, the conclusion is that the new p-GTO device exhibits superior switching characteristics compared with the classic GTO. Recent works [9,10] show that implementation of similar devices on Silicon Carbide (SiC) translates into lowering the turn-off times up to 10 times.

Also, in the simulation was computed the static operating point of the devices, i.e., the anode current I_{AT} = 400A and the on-state voltage V_T. In both cases the obtained value was V_T = 1.5 volts.

7. Conclusions

The p-GTO (gate-turn-off) thyristor has a p-type low-doped base, in comparison with its classic counterpart, which has a n-type low-doped base. As shown in this work, design optimization techniques applied with SILVACO TCAD simulation tools lead to improved static and dynamic characteristics of the p-GTO thyristors. The main advantages are augmented breakdown voltages of the power device, improving the main anode-to-cathode maximum voltage and the cathode-to-gate breakdown voltage, leading to the possibility of faster turn-off switching.

8. References

[1] Silard A P 1990 *Gate Turn-off Thyristors GTO – Functions, Principles, Structures, Applications, (in Romanian)* (Bucharest: Technical Publishing House)

[2] Laughton M A and Warne D F 2003 *Electrical Engineer's Reference Book, 16th ed.* (Elsevier Science) pp 17/24-5

[3] Wilamowski B M and Irwin J D 2017 *Power Electronics and Motor Drives 2nd ed.* (CRC Press, Taylor & Francis Group) pp 1-9-10

[4] Silard A P and Duta M J 1992 The P-GTO thyristor - a novel power switching device *International Seminar on Power Semiconductors - ISPS '92, Prague*
[5] Silvaco Inc. 2013, *ATLAS User's Manual*
[6] Silvaco Inc. 2013, *ATHENA User's Manual*
[7] Silvaco Inc. 2013, *DeckBuild User's Manual*
[8] Silvaco Inc. 2014, *TonyPlot User's Manual*
[9] Wang J, Wang G, Li J, Huang A Q, Melcher J and Atcitty S 2008 Silicon carbide emitter turn-off thyristor *Intl. J. Power Mgmt. Electronics*, *Hindawi Publishing Corporation*, Article ID 891027, doi:10.1155/2008/891027
[10] Ryu S H 2018 Blocking performance improvements for 4H-SiC p-GTO thyristors with carrier lifetime enhancement *Materials Sci. Forum* **924** pp 633–636 https://doi.org/10.4028/www.scientific.net/msf.924.633

Analysis of methods for assessing the energy efficiency of data centers using the power usage effectiveness method

R Uanov[1], A S Begimbetova[2]

[1] Satbayev University, Almaty, Republic of Kazakhstan
[2] Almaty University of Energy and Communications named after Gumarbek Daukeev, Almaty, Republic of Kazakhstan

E-mail: a.begimbetova@aues.kz

Abstract. The article deals with the analysis of methods for assessing the energy efficiency of data centers according to the Power Usage Effectiveness method. The demand for data centers which consumes a large amount of electricity is growing with the growth of digitalization and the accumulation of big data in the network. The energy consumption of the cooling system for the machine room accounts for a significant part of the operating costs of the building. Free cooling in a refrigeration system reduces energy consumption much more than operating systems with a vapor-compression cycle. In 2006 according to The Green Grid, the assessment method of Power Usage Effectiveness has become an international standard for measuring energy efficiency and is widely used in the design and operation of data centers. In this regard, the operation principles of free-cooling chillers are considered. The calculation example of the system payback in free-cooling is also given.

1. Introduction

Currently, the global economy makes extensive use of information and communication technologies (ICT) and related technologies for generating, transmitting, distributing, processing and storing digital data. All markets are seeing exponential growth in the volume of such data in the social sphere, in the field of education and business, with the Internet acting as the backbone transport network, and many private companies and other organizations operate their own data centers and leased facilities, acting as nodes and hubs. The generation of data and the subsequent increase in the workloads associated with the processing and storage of this data are directly related to the increase in power consumption.

The data center plays an important role in calculating, storing and managing digital information on the Internet. The expansion of the data center and, in particular, the increase in electricity consumption is an inevitable consequence of these processes, and will require an increase in electricity consumption, despite the most stringent energy efficiency strategies. This necessitates KPIs that measure resource efficiency and deliver significant emissions reductions. The data center function is focused on solving business problems by providing information services and is the operator of many processes: from ordinary correspondence in the messenger to the implementation of an online banking transaction. In the Republic of Kazakhstan, the activity of a data center is regulated by standards ST RK ANSI/TIA-942-A-2016; ST RK DSP-IS 0101_1.0.0; ST RK DSP-IS 0102_1.0.0.

Today, the main problem of energy consumption in a data center is the functioning of the ventilation and air conditioning system. Since the network and server computing equipment located in the computer room of the building, when processing and calculating information in the 24/7/365 mode, heats up and is a source of heat surplus. Solving the problem of reducing power consumption by refrigeration systems is very important, since according to the global industry, a rapidly growing data center will consume 4.5% of the electricity generated worldwide by 2025. For comparison, in 2010 this figure ranged from 1.1% to 1.5% (325 billion kWh).

Generation of cold, obtained in a vapor compression cycle based on the processes - compression, condensation, throttling and evaporation, is a very unprofitable measure in terms of energy consumption, since compressors in a year-round operation system consume a large amount of electricity, usually up to 37% of the total electricity consumed in a data center ... Until recently, data centers were energy-consuming facilities, since only vapor compression refrigeration systems were used to cool the turbine hall, which consume excessive amounts of electricity [1-2].

2. Objects and research methods

A computer center (data processing center) is an intelligent building that houses server and telecommunications network equipment for processing multiple digital data.

Unlike conventional server computing rooms, in the data center all service engineering systems are reserved, divided according to the degree of fault tolerance, and besides, the systems are continuously monitored to prevent emergency situations or promptly respond to them.

It should be noted that not all of the electricity consumed ensures the operation of IT equipment, for example, 50% of electricity (figure 1) is consumed by engineering infrastructures: refrigeration systems, ventilation and peripheral systems.

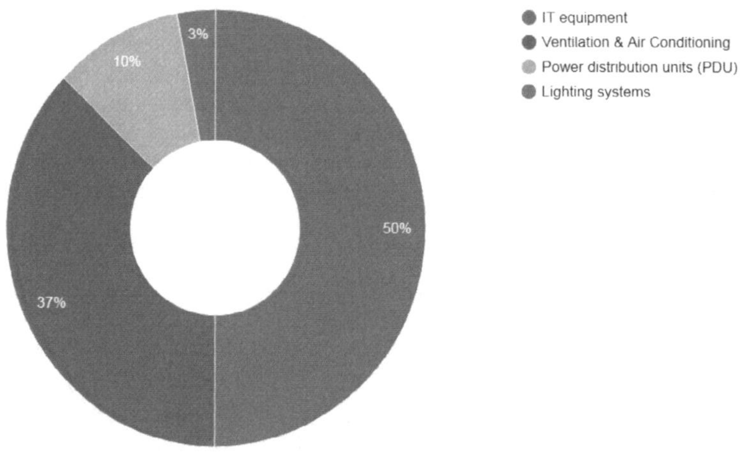

Figure 1. Electricity consumption in the data center as a percentage

Competent monitoring of electricity consumption in a data center can improve its efficiency. The standard metric for data center efficiency is the energy efficiency ratio (PUE) proposed by the international organization Green Grid in 2007. With the advent of the "green energy" trend in the data center industry, the Power Usage Effectiveness (PUE) has been applied to assess environmental friendliness and energy efficiency. This coefficient is the ratio of the total power consumed by the data center to the power consumed by the IT equipment. According to Green Grid, the energy utilization factor allows [1,4]:

- determine the possibility of improving the efficiency of the data center;
- compare the performance of one data center with others;
- determine the possibility of expansion and redesign of the data center in the future;
- determine the possibility of reorganizing the power supply to connect additional IT equipment;
- determine the calculated capacity values when designing a new data center.

The PUE formula is as follows:

$$PUE=\frac{Total\ Power\ Consumption\ of\ the\ Data\ Center}{Energy\ consumption\ of\ IT\ equipment} \quad (1)$$

The calculation of the PUE ratio is based on dividing the electricity consumed by the data center infrastructure as a whole by the energy consumption of IT equipment. Determining the energy efficiency of data center infrastructure will help us take the first step towards reducing overall energy consumption and energy costs. It allows you to assess the current layout of a particular data center and develop an optimal strategy for further development.

Determining the PUE provides useful information and guidance for designing efficient power and cooling architectures, deploying equipment within those architectures, and operating that equipment.

PUE helps determine:

- opportunities to improve operational efficiency in the data center;
- the efficiency of optimization of structures and processes in the data center in dynamics;
- design goals for new data centers, taking into account the full range of expected IT workloads.

It is important to note that the lower the PUE value (figure 2), the more energy efficient the data center operation is considered. The ideal ratio is close to 1.0. If the average PUE per year is, for example, 2.5 units, this means that the data center as a whole uses 2.5 times more energy than all IT equipment in general (including servers, storage systems and active network equipment) [1, 3-4].

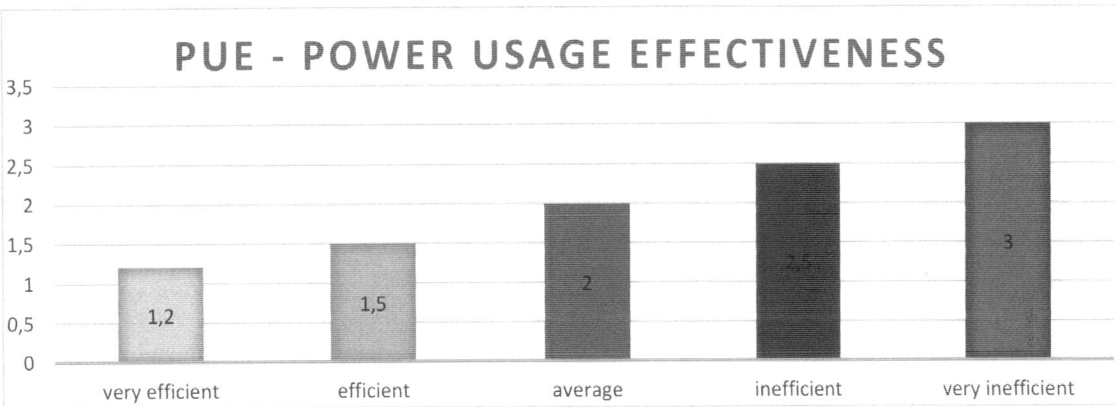

Figure 2. PUE value. Energy efficiency indicators

Reducing PUE is mainly achieved by installing a free-cooling system. The principle of freecooling is based on air heat exchange in a heat exchanger without mixing the outside and inside air. During the summer period, water is used as a coolant; in winter, a water-glycol antifreeze solution is used. A combined version is also allowed. The use of free-cooling allows you to shift the task of heat transfer only to fans, which consume much less electricity. If these fans are equipped with energy efficient EC (Electronically Commutated) motors with variable speed, then energy costs are reduced even more.

Today, the following types of freecooling are used:

- chiller with free cooling;
- direct free cooling;
- indirect free cooling with or without adiabatic humidification of the outside air.

Chillers by installation type are:

- monoblock (figure 3 a, b);
- with a remote air condenser (figure 4 a, b);
- with a water condenser based on a dry cooler (figure 5 a, b);
- combined with free-cooling function.

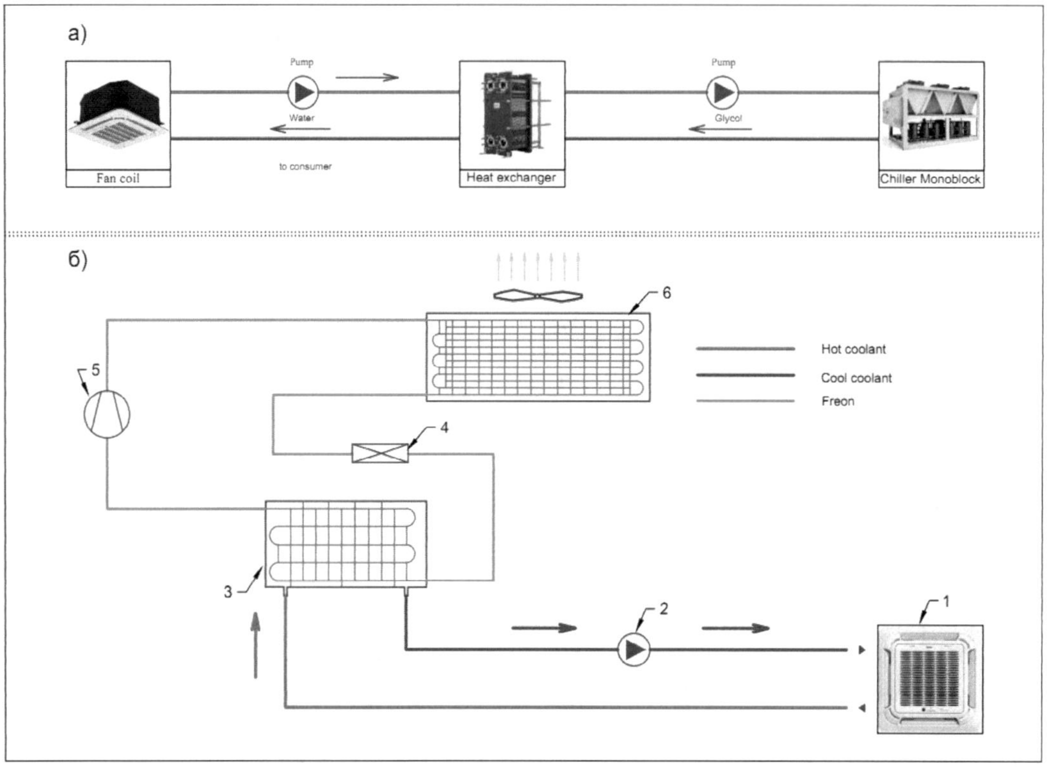

Figure 3. Typical circuit solutions of a monoblock chiller: 1 - fan coil; 2 - coolant pump; 3 - evaporator; 4 - throttling device; 5 - compressor; 6 - air condenser.

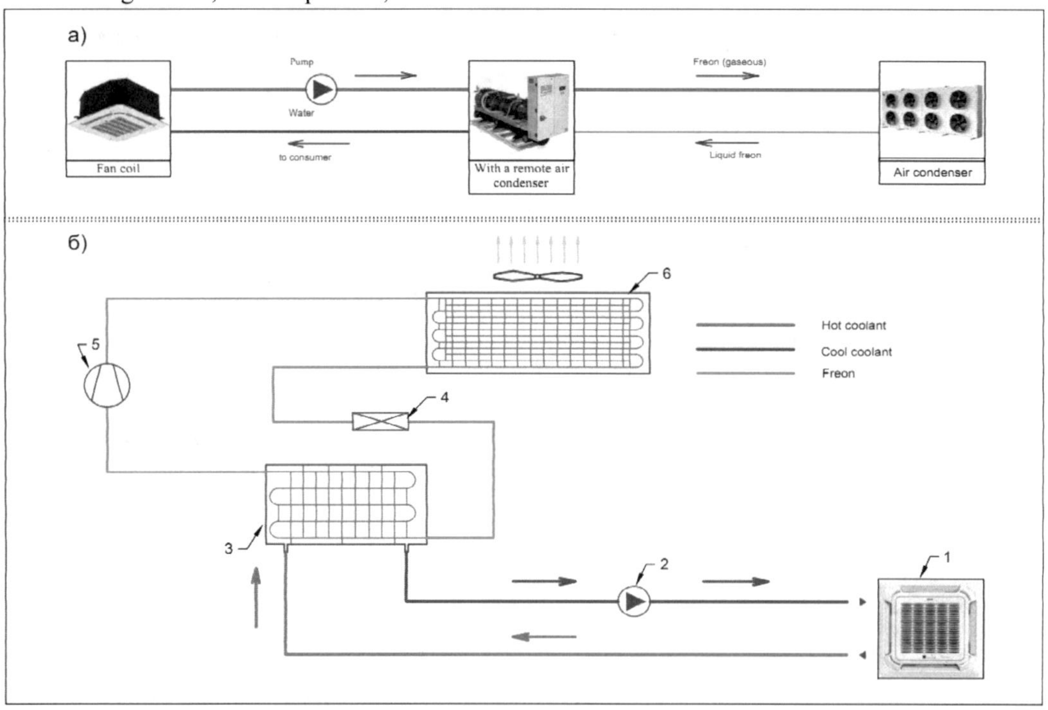

Figure 4. Typical circuit solutions of a chiller with a remote air condenser: 1 - fan coil; 2 - coolant pump; 3 - evaporator; 4 - throttling device; 5 - compressor; 6 - remote air condenser.

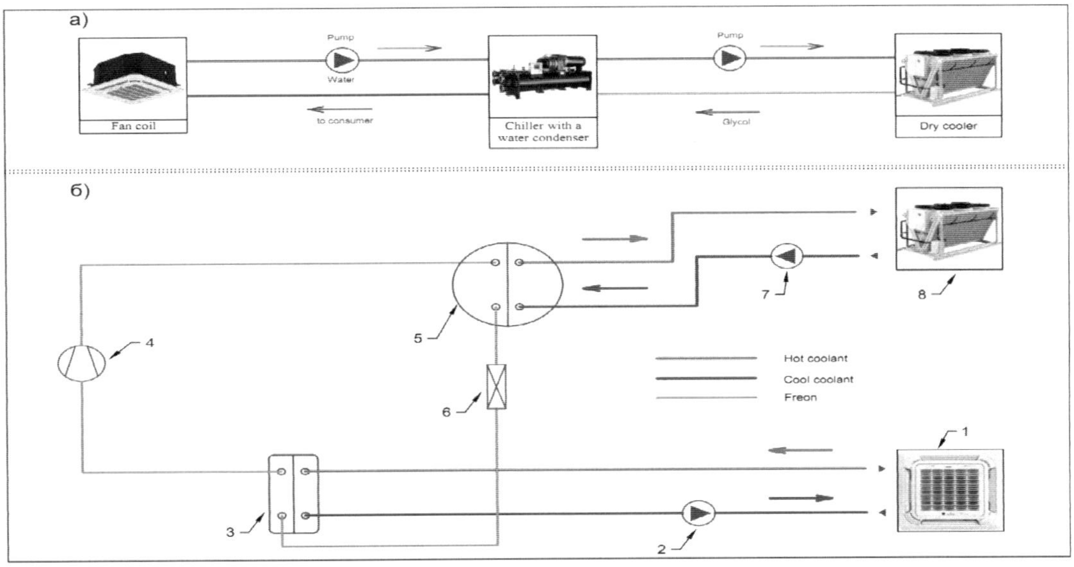

Figure 5. Diagram of a chiller with a water condenser: 1 - fan coil; 2 - coolant pump; 3 - evaporator; 4 - compressor; 5 - water condenser; 6 - throttling device; 7 - condenser coolant pump; 8 - dry cooler.

The energy consumption for the operation of cooling systems depends on the ambient temperature. Energy consumption measurements taken during the cold season, when the air conditioning systems can be turned off and cooling is provided by the cold outside air, give a PUE value closer to unity than measurements taken during the hot season.

When the outside temperature allows the temperature of the cooling liquid to drop below the required temperature of the cooling medium, the compressor in the freon circuit is switched off and the three-way valve redirects the cooling liquid to the free-cooling heat exchanger. Thus, the cooling of the coolant in the freecooling heat exchanger occurs due to the low temperature of the outside air.

The chiller circuit has a water cooled condenser. Refrigerant condensation occurs due to the removal of liquid, which, in turn, is cooled in a dry cooler (dry cooler) and has two circuits - fren and water solution (or water-glycol antifreeze solution).

Freecooling chillers are larger, heavier and more expensive than conventional vapor compression chillers. However, the high energy efficiency of a freecooling chiller is achieved by shutting down the steam compressors during the cold season and benefits from energy savings at the end of the payback period. Since the steam compressor consumes up to 90% of the total electricity consumption of the entire system.

For chillers with a free cooling system, several operating modes can be distinguished.

One of them is the machine cooling mode. When the outside temperature does not allow the coolant temperature to drop below the required coolant temperature ($t_{coolant} > t_{refrigerant}$). In this case, natural cooling of the coolant is physically impossible. In this mode, the chiller operates like a conventional chiller with a water condenser in a freon circuit (Figure 6), a three-way valve redirects the coolant to the chiller condenser. Cooling of the refrigerant from the consumer takes place in the evaporator of the chiller, due to the removal of heat to the refrigerant. As a result, the refrigerant boils away, the refrigerant vapors are sucked into the compressor, then the compressor compresses the refrigerant vapor to high pressure and a temperature higher than the temperature of the coolant. After that, refrigerant vapors are pumped into the condenser. In the condenser, the refrigerant is condensed by transferring heat from the refrigerant to the coolant. The coolant is in turn cooled in the dry cooler by heat exchange with the ambient air. The liquid refrigerant from the condenser decreases its pressure, passing through the throttling devices and again enters the evaporator. This cooling cycle is repeated. The circulation of the coolant and the coolant is carried out using pumps.

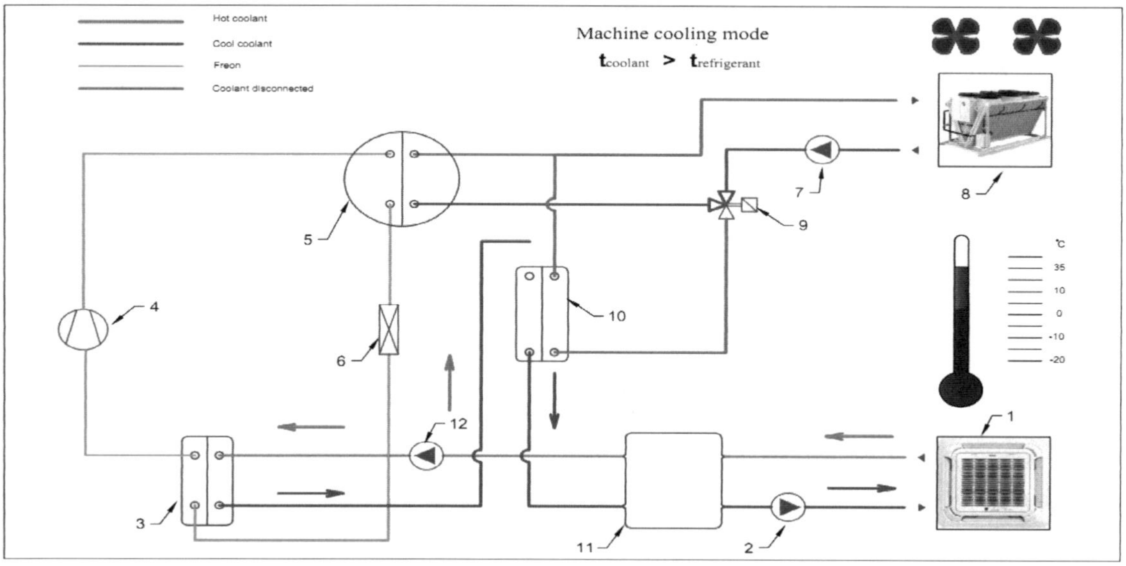

Figure 6. Machine cooling mode.

A transient mode is a process when the outside air temperature ensures that the coolant temperature drops below the required coolant temperature *($t_{cooling\ liquid} < t_{coolant}$)*, but at the same time 100% of the cooling capacity is not yet ensured due to free cooling (figure 7).

In this case, the three-way valve is in an intermediate position. Part of the coolant is directed to the condenser to ensure the operation of the freon circuit, the other part is directed to the free cooling heat exchanger. In this mode, the coolant is cooled in two stages: first in the evaporator due to the operation of the freon circuit, and then in the freecooling heat exchanger due to heat exchange with the coolant from the dry cooler. The compressor operates in this mode, however, the power consumption is significantly reduced, since due to the operation of the freon circuit, the heat removal is reduced. Compressors stop periodically. If the chiller is multi-compressor, then some of the compressors stop, as a result of which the power consumption is reduced.

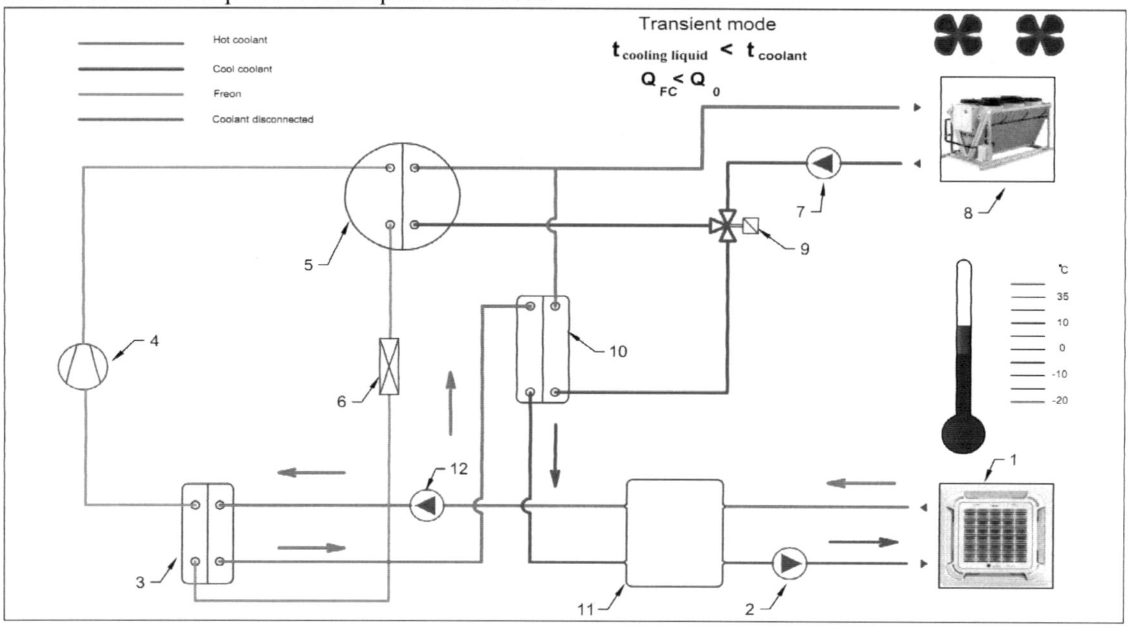

Figure 7. Transient mode.

Free-Cooling occurs when the outdoor temperature drops the coolant temperature below the desired coolant temperature (t_{cool} <$t_{cooling}$), 100% cooling capacity is achieved by free cooling and the chiller compressor stops. The system goes into freecooling mode.

Cooling of the coolant occurs due to heat exchange with the coolant from the dry cooler. In this mode, the three-way valve completely redirects the coolant to the free cooling heat exchanger. The freon circuit does not work in this case. Maximum energy savings observed in freecooling mode [9-10].

3. Results of the study

At the design stage of a free cooling system in a data center, one of the main parameters for choosing an energy-efficient solution is the payback period of this project and the value of the annual energy savings. To estimate the payback period, the formula is used:

$$\tau = \frac{CC_2 - CC_1}{(N_{el.1} - N_{el2}) * n * T} \quad , \tag{2}$$

where τ – payback period of capital expenditures increase, years; Ŧ – annual savings on electricity, tenge; CC_1 – capital costs for equipment without FC, tenge (KZT);CC_1 – capital costs for equipment with FC, tenge (KZT).

The amount of annual energy savings is determined by:

$$Ŧ = (N_{el.1} - N_{el.2}) * n * T, \tag{3}$$

where $N_{el.1}$ – power consumption of equipment in machine cooling mode, kW;$N_{el.2}$ – power consumption of equipment in FC mode, kW; n – annual operating time of equipment in the FC mode, h; T – electricity tariff, tenge (KZT).

As an example, the calculation of the payback of a freecooling system for installation in a computer room (figure 8), consisting of 4 rows of 13 server racks each, 52 server racks in total, is given. The machine room is located in the city of Almaty. Each rack contains 15 servers with an average power of 400 watts. Since the heat dissipation of IT equipment is equal to its power consumption, the heat surplus of each rack is 600 W = 6 kW; then the knowledge of the total heat dissipation of the equipment will be 312 kW.

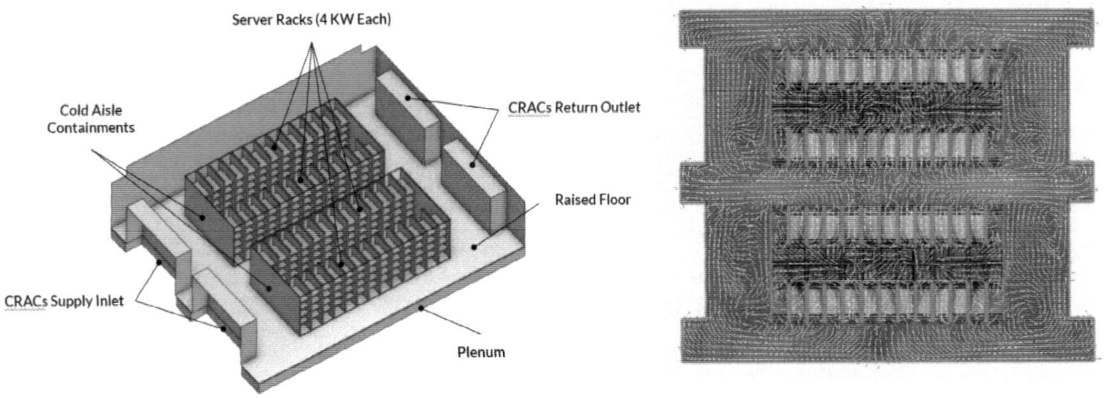

Figure 8. General diagram of the turbine room [12].

Required cooling capacity 312 kW. A monoblock chiller with and without FC system was taken as a refrigeration machine.

CO_1 → Commercial offer (229 305,00 € * 509,20 KZT = 116 762 106,00 KZT);
CO2 → Commercial offer (174 054,00 € * 509,20 KZT = 88 628 296,80 KZT);
Nel.1 → Chiller specifications without FC (131,4 кВт);
Nel.2 → Chiller Specifications with FC (12 кВт);

n → The set of rules of the Republic of Kazakhstan 2.04-01-2017 "Construction climatology" (179 days / at a temperature of 0.8 0C / * 24 hours = 4296 hours) [13];

T → Average selling rate for legal entities – 20,69 KZT with VAT [14].

Dollar to tenge rate: 1USD = 509.20 KZT [15].

$$₸ = (131.4 - 12) * 4296 * 20{,}69 = 10\ 612\ 778{,}26 \text{ KZT} \quad (4)$$

$$\tau = \frac{116762106{,}00 - 88628296{,}80}{(131{,}4-12)*4296*20{,}69} \approx 2{,}7 \quad (5)$$

The payback period for a monoblock chiller with free cooling for Almaty is ≈ 2.7 years.

In regions with temperate and cold climates, the availability of free-cooling helps to significantly save the amount of electricity in the autumn-winter period. In a dry cooler, only axial fans are powered, while in other types of refrigeration units, electricity is also required to operate the evaporator.

4. Conclusion

To improve the energy efficiency of the data center, free cooling systems should be introduced at the design stage. When implementing, planning, building a data center to improve the PUE coefficient, one should take into account the climatic characteristics of the environment, the possibility of using renewable energy sources and the rejection of the energy-consuming vapor compression refrigeration cycle in favor of more efficient cooling systems based on chiller free cooling.

Free cooling systems used in data centers are more costly to install than a vapor compression system, but the capital investment pays off during their operation. At the same time, the development of energy efficient technologies does not stand in one place. To achieve energy efficiency, cold seawater, groundwater, etc. can be used as a chiller. The data center cooling system, built on free cooling, not only works more stable, but also allows you to achieve the average annual PUE that is several times lower, without the cost of expensive energy-efficient components (chillers, cooling towers). In this case, when calculating the PUE, not only the ratio of the IT load is taken into account, but also the energy consumption for the refrigeration system.

5. References

[1] Liu Y, Wei X, Xiao J, Liu Z, Xu Y and Tian Y 2020 Energy consumption and emission mitigation prediction based on data center traffic and PUE for global data centers *Global Energy Interconnection* **3**(2) pp 272-282 https://doi.org/10.1016/j.gloei.2020.07.008

[2] RK ANSI/TIA-942-A-2016 Telecommunication infrastructure of data centers (ANSI/TIA-942-A-2012 Telecommunications infrastructure standard for data centers, IDT)

[3] Lebedev N N 2018 Eatures of using" green "energy for dpc needs *Journal of Herald of Contemporary Research*, 9.3 (24) pp 283-284

[4] PUE™: A comprehensive examination of the metric https://www.thegreengrid.org/

[5] Altoona Data Center. https://web.facebook.com/AltoonaDataCenter [date of appeal 23.10.2020]

[6] Kirillov I, Kovlenko K 2015 Effective data center about PUE and more *Journal Networks and Business* **80** (1)

[7] Oganesyan E S 2014 Using PUE as an energy efficiency indicator in data centers. *Journal: Ecology of Industrial Production*, pp 77-82

[8] Chillers with free cooling system and water condenser. Principle of operation. Payback period. https://lessar.com/support/learning/

[9] Diagram of the free cooling system of a monoblock chiller. https://lessar.com

[10] How do adiabatic cooling systems work? https://www.waterchillers.com

[11] Data Center Cooling Cost Reduction - Editorial Demo' simulation project by vaibhav_s. Public project: https://www.simscale.com

[12] Electricity tariff for Almaty. https://esalmaty.kz/ru/business-tariffs

[13] Beauregard M A and Ayer S K 2018 Maintaining performance: Understanding the relationship between facility management and academic performance at K-12 schools in the State of Arizona *Facilities*, 36 pp 618-634 https://doi.org/10.1108/F-11-2017-0111
[14] Bieser J and Menzel K 2019 Assessing Facility Maintenance Models for Data Centres: Status and Deficits of Current Facility Management and Maintenance Concepts, *Applied Mechanics and Materials* **887**, pp 255–263 https://doi.org/10.4028/www.scientific.net/AMM.887.255
[15] Braglia M, Castellano D and Gallo M 2019 A novel operational approach to equipment maintenance: TPM and RCM jointly at work, *Journal of Quality in Maintenance Engineering*, **25** (4), pp 612-634 https://doi.org/10.1108/JQME-05-2016-0018
[16] Abbas C J B Orozco A L S and Villalba L J G 2015 Monitoring of Data Centers using Wireless Sensor Networks, *Handbook on Data Centers* pp 1171–1183 doi: 10.1007/978-1-4939-2092-1_40
[17] Nor N M, Hussin M and Abdullah R 2019 Energy-saving framework for data center from reduce, reuse and recycle perspectives. *Pertanika Journal of Science & Technology* **27**(3) 1259-1277
[18] ISO 26382 Cogeneration systems - Technical declarations for planning, evaluation and procurement https://www.iso.org/standard/43552.html
[19] ISO/IEC 20000-1 Information technology - Service management - Part 1: Service management system requirements https://www.iso.org/standard/70636.html
[20] IEC 62052 (series) Electricity metering equipment (AC) - General requirements, tests and test conditions.

Models and methods of digital mechanisms in anti-corruption, their advantages and disadvantages, and applications

A Nurkey[1], A Mukasheva[2] and D Yedilkhan[3]

[1] Academy of the Public Administration under the President of the Republic of Kazakhstan, Nur-Sultan, 010000, Kazakhstan
[2] Almaty University of Power Engineering and Telecommunications, Almaty, 050013, Kazakhstan
[3] Astana IT University, Nur-Sultan, 010000, Kazakhstan

E-mail: a.nurkey@gmail.com

Abstract. Corruption is one of the main problems in many developing countries. However, the complexity of measuring corruption and its consequences does not allow for its complete study and implementation of measures. The factors and indicators currently known worldwide cannot measure corruption on time scales and depend on a narrow circle of experts in this area. Thus, corruption is easily confused with institutional gaps. In modern society, where the technologies such as Data Science and Predictive Analytics play a huge role, corruption is still omnipresent. The article examines the priority areas of combating corruption using new digital technologies. The main direction of the article is defined as an analysis of the advantages and disadvantages of the digitalization in the areas of solving social conflicts. The article presents the comparative analysis of technologies of digital anti-corruption compliance in developing countries, on the example of Kazakhstan. At the same time, according to the results, the article discusses the disadvantages of using proposed models due to the peculiarities of the legislation.

1. Introduction

Information and communication technologies (ICT) in general play an essential role in many areas of management, including government, through the transparency of the business processes. It can be said, that ICT technologies provide more efficient and fair governance. This paper shows the attitude of civil society towards ICT as both an anti-corruption tool and a governance tool. The use of ICT in the fight against corruption is gaining more and more popularity worldwide. The desire to use modern approaches to counter corruption, fueled by the acceleration of ICT development, and has led to new exciting solutions.

Nowadays, ICT can be actively used to ensure integrity in the provision of public services, in strengthening the public reporting system, public procurement, and public financial management. The purpose of the paper is to study the impact of ICT in combating corruption and analyze the economic benefits of implementing modern technologies in the anti-corruption policy on a regional level. In addition, there will be points examining the relationship between factors related to technologies and corruption perceptions. From the technical side, we should define the government policy and vision related to technology, implementation of technological tools in legislative documents of corruption prevention, and its economic and social efficiency in the public sector.

This research identifies the relationship of the above factors with corruption. Moreover, this paper examines the positive influences of the public policies and the capability of the digital legal system to digital maturity.

Hypothesis 1: Digitalization of public services positively influences anti-corruption policies. The free access to the technologies makes people combine their business and household with the internet, requiring high quality of connection and cell phone communications. Easy access to the internet, its low price, good quality of connection, and widespread popularity may lead to developing e-commerce. To make transactions through the internet, users need to upload customers' personal information like ID number of credit card, security codes, and others. However, the lack of security could negatively relate to the number of users online. One of the ways to avoid this is to implement high-quality IT and internet infrastructures.

Hypothesis 2: ICT positively influences the trust of citizens in the government. The paper also questions the SWOT analysis of using ICT that should have been chosen to fight corruption effectively.

The article consists of Introduction, Corruption, Corruption in Kazakhstan, where the article's literature review, aims, and hypothesis can be found. Then a Research method and Analysis section is provided, and after that, the research results are given in the section Discussion and conclusion.

2. Corruption in Kazakhstan

Since gaining independence, Kazakhstan has not ceased its fight against corruption, and during this relatively short period, conditional stages in the development of anti-corruption strategies and activities, in general, can be distinguished.

The first stage begins with implementing the Law of the Republic of Kazakhstan dated July 2, 1998, No. 267-I "On the fight against corruption" [3]. At this stage, the country was actively fighting the consequences and sources of everyday corruption since, after the collapse of the Soviet Union, all CIS countries faced this phenomenon. Citizens had to deal with the facts of corruption when receiving public services, whether in medical institutions or law enforcement agencies. At this stage, the state had to minimize the contact of civil servants with service recipients. Thus, adopting the advanced experience of developed countries such as Singapore, Great Britain, France, and others, one-stop shops appeared in the country, operating on the principle of a "single window".

The second stage of development began in 2010-2015 when the state introduced the Law of the Republic of Kazakhstan dated November 18, 2015, No. 410-V "On Combating Corruption" [3]. The work with the principle of "single window" is gaining momentum. Therefore, further steps are determined in business corruption and work with artificial barriers in the private sector. Some works defined that the private sector and corruption prevention are correlated [1]. Corruption in the quasi-public sector is suppressed, the criminal code is being hardened in the direction of receiving and giving bribes. The state has a comprehensive approach in the fight against corruption.

Some of the most critical corruption factors in Kazakhstan are: legal, organizational and managerial, economic, socio-cultural, and psychological [2].

From the point of view of the legal support of this issue in Kazakhstan, there is no targeted anti-corruption document regulating the work of government officials and civil society. Also, one cannot fail to consider that the legislative bodies are not entirely removed from bribery. There is still the possibility of active participants in this crime among the employees of state bodies. In this regard, it is necessary to ensure modern and working laws. It is fashionable to include supply and demand for bribery among economic factors. According to economic realities, where there is a demand for corruption, an offer will undoubtedly appear. Also, in some cases, a monopoly on corruption appears. The issuance of public service by one state body is monopolized, thereby creating conditions for the monopolization of this public service. Kazakhstan is one of the former CIS countries where the corruption level slightly differs from the others in 2019, which can be seen in figure 1[4].

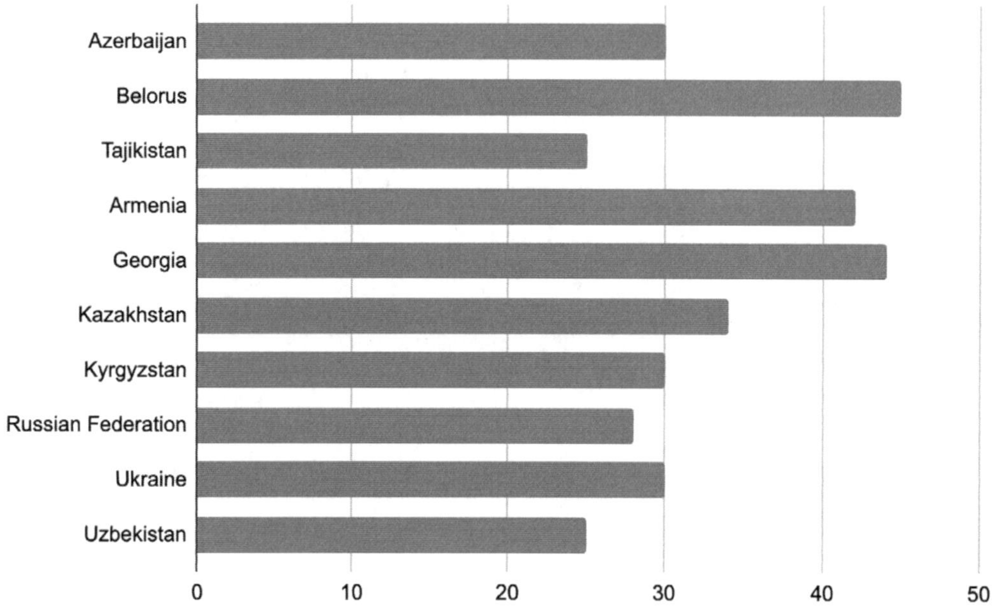

Figure 1. Corruption Perceptions Index of Post-Soviet Countries for 2019.

The Government has a clear vision of trade and commerce prosperity as well as citizens' wellbeing. Therefore, it has to implement new applications and technologies to achieve the purpose.

3. Corruption and ICT

New technologies are facilitating simplified work between the government and its citizens. The main advantages of ICTs are that they can be used to increase the transparency of processes, the level of accountability, and the quality of information dissemination [5]. Other scholars confirm this assumption [6] and believe that the use of ICTs contributes to the automation of main business processes, limits the actions of some unethical civil servants, and reduces red tape. However, they cannot fail to note the fears regarding using this technology as a tool for committing this crime. Open data and access to the rapid development of the Internet and technologies can contribute to hacking all kinds of sites and databases, distorting or hiding some information. Kossow and Dykes investigated the concept of the benefits and harms of ICT in countering corruption and concluded that this would depend on the impact on the demand or supply of certain information [7].

So, after analyzing literature review, we can identify eight main factors: automation, transparency, monitoring, informatization, accountability, prevention, encouragement, and training. Nowadays the popularity of using neural networks are expanding not only in science but also in private sectors [8]. These practice shows that the artificial intelligence could be successfully used in public sector. The corruption manifestations could be diminished by these tools.

Below table 1 can be seen, where the central area of using ICT in corruption prevention is constructed. In the table, we tried to show the main directions of where ICT can help to minimize corruption manifestation in any possible way.

Table 1. The types of using ICT in fighting corruption

Action type	Actions	Main goal	ICT
Automation	Eliminate human impact on the process	Fight with petty corruption	Automated solutions and information systems, Big data and predictive analytics, Blockchain, e-government
Transparency	Eliminating the possibility of hiding information	Mobilizing citizens, service recipients	User-friendly websites where all information is published, Blockchain, Big data, Open Data portals
Monitoring	Monitoring the subject of identifying preparation for committing corruption	Fight with large scale corruption, for example in public procurement and foreign trade	Social media tools and analysis, Big data, Blockchain, cross validation, Artificial intelligence
Informatization	Notifying Citizens of Government Rules and Procedures to Reduce Resistance to Covering Corruption	Fight with petty corruption	Any information systems, user friendly websites, cloud technologies, usage of QR codes
Accountability	Mobilize users and communities to disclose cases of corruption (Whistleblower)	Fight with petty corruption	Any information systems, social networks, e-government, blockchain, Artificial intelligence
Prevention	Publication on social media and other sources of information and a report on corruption	Fight with petty corruption	Social networks, double audit, Blockchain, Artificial intelligence
Encouragement	Organization of consultation meetings, conferences online	Changing the behavior of citizens	Social networks, blockchain, QR code (via gaining social encouragement)
Training	Train government officials in order to widespread knowledge about corruption prevention actions and mechanisms	Changing the behavior of citizens and civil servants, enhance trust of the citizens	Trainings, meetings, QR code, Cloud technologies, social networks

4. Research methods and analysis

The research consists of questionnaires of 118 respondents of the citizens of Nur-Sultan city in Kazakhstan in order to analyze the effectiveness of ICT and get their typical attitude and expectations toward the topic.

One of the questions was "Which of the parties, in your opinion, is more interested in committing the act of corruption?" (Figure 2), where respondents share opinions regarding the most exciting side in committing corruption. The results show that the most interested player is acceptors. That could illustrate that the ICT tools must be implemented in order to limit the ways of accepting corruption rather than giving it. For example, through Blockchain, Big data, and Cloud technologies, which can help to provide transparent business-processes.

IOP Conf. Series: Materials Science and Engineering 1216 (2022) 012015 doi:10.1088/1757-899X/1216/1/012015

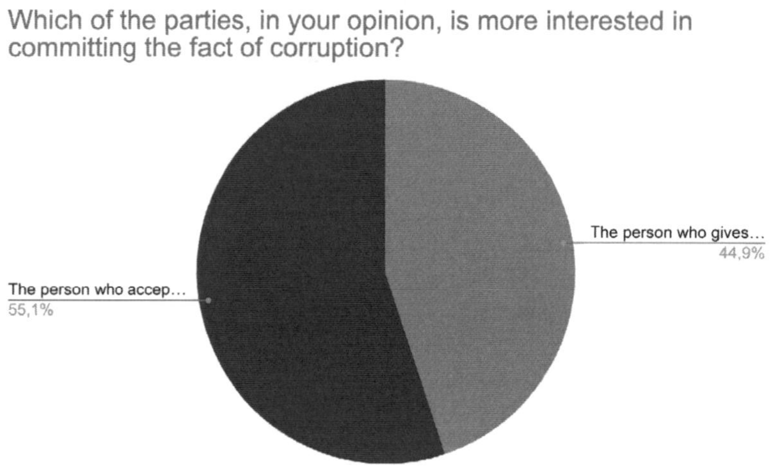

Figure 2. Analysis of question "Which of the parties, in your opinion, is more interested in committing the act of corruption?".

Almost 87 percent of respondents have confidence that ICT benefits corruption prevention. However, there are still 7.6 percent skeptical citizens, who have negative thoughts about the initiative (figure 3). As we can see the vast proportion of the respondents have a strong believes in ICT and its benefits in the field of curbing bribery.

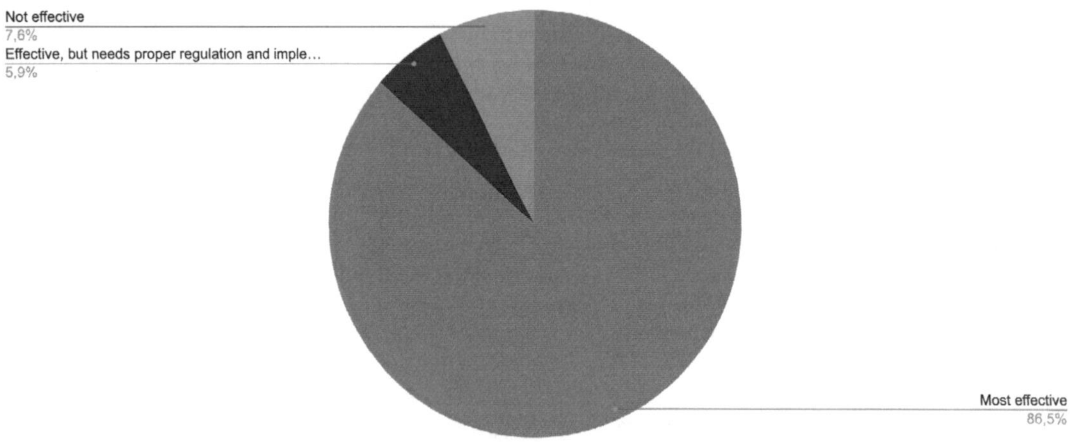

Figure 3. Analysis of question "How effective are new technologies such as artificial intelligence, distributed systems (Blockchain) and Big Data for fighting corruption?".

Trust of the citizens to the government has a significant impact on corruption prevention because this will enhance the whistle-blowing mechanism in general and help distribute the information regarding it. In this case, corruption perception through the citizen's trust in the government was assessed (figure 4). The survey shows that the degree of trust in the government's anti-corruption policies in Kazakhstan is shallow. This can be because citizens do not have strict insurance, and they could have experienced or heard about the dishonest actions of government officials.

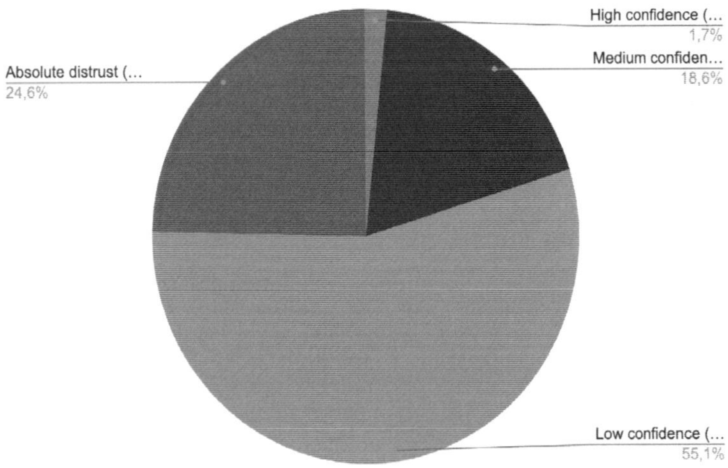

Figure 4. Analysis of question "The degree of your confidence in the state and government reforms in the field of detection of corruption?".

The SWOT analysis of different types of ICT (Table 2) was performed, where implementation of some of ICT technologies (Blockchain, Cloud technologies) were analyzed and both the effectiveness and disadvantages were described. For example, Fauvel in his works describe the economically ineffective of blockchains, because of a huge energy consumption, however this can be the most useful tool in monitoring corruption and shadow economy [9]. Cloud Technology has a lot of advantages, although the security part is not so safe [10]. The main problem of QR codes is that they may be distorted which can lead to the distribution of damaged information [11]. This kind of obstacles are common for all types of Information Communication technologies [12, 13].

Table 2. SWOT analysis of different types of ICT in corruption perception.

Types of technology or tools	**Strengths**	**Weaknesses**	**Opportunity**	**Threads**
QR code (Quick response code)	Versatile User-friendly Fast scanning Economically effective Storage advantages Eco-friendly Business-friendly	Poor execution Not famous among elderly people Need a smartphone	Helps to fast inform a huge amount of people concerning corruption Helps to storage some necessary information and materials	Security aspect is not so strength Those, who don't have a smartphone could miss information
Blockchain	Process integrity Security Detectable	Not eco-friendly Economically ineffective Not power-consuming High costs of implementation and operation	Reduce potential corruption risks Provide strong accountability Reveals shadow economy and corruption in the system	Consume huge energy and money, and this could reduce the benefit itself Needs high qualified IT specialists

Cloud technology	Easy implementation Accessibility Economically effective High speed Mobility API Access availability	Bandwidth issues Technical problems Depends on Internet connection	Provide unlimited storage opportunity Web-based control	Security Threat Performance is not stable Economically not so effective (because of operational costs)
Big Data	The availability of data from different sources Accessibility Economically effective Forecast Open data opportunities	Data requires a lot of processing Data storage format requires unification	Provision of open data Ability to use API to connect third-party applications Possibility of forecasting	Data leakage threats Data processing problems
Artificial intelligence	Accurate prediction Simplified results interpretation Constant monitoring	Lack of data Limitation in performance	Development of novel techniques Continuous implementing of new data	Policy and legislation Personal data abuse Ethical questions

Some ICT technologies are not so safety as we can think, there could be some threats related to physical damage, loss of information and cyber security issues. These threats have a huge impact on the liquidity and capital in general [12]. The cost should be calculated before applying the ICT technologies.

5. Discussion and conclusion

New technology certainly brings a lot of advantages in life of ordinary people as well as in the work of government. The survey shows that trust of the citizens is low and they are not confident about corruption prevention mechanisms which are implemented in Kazakhstan. However, according the same survey they have a positive attitude to the issue of ICT in fighting corruption. Almost 9 out of 10 respondents have a believes in the ICT in public sector. The respondents have a slightly more opinion that bribe receivers are more motivated than givers. That is why, we can consider that the most effective for enhancing public trust may be a monitoring and prevention works with public officials. That could give us the positive results of hypothesis 2.

In the paper we also consider the SWOT analysis of the types of ICT that should have a precise attention of government officials. So, at the end we can see that some of the tools have absolutely advantages to use in public sector, like QR code and Artificial intelligence. However, some of them have a difficulties and effectiveness threads like, economic and security. These tools (Blockchain, Cloud technologies and Big Data) needs additional resources or updates to be effectively useful in public administration. As a result, the hypothesis 1 could not be absolutely right in our case.

As we can see from the analysis, the citizens of Kazakhstan are positively reacting to a new initiative which is concluded in implementing new technologies in the governmental sectors. Although currently, they are not sure about the effectiveness of the working policy and overall, to the government.

Government of Kazakhstan should implement the ICT tools in corruption prevention policies although the outcome should be calculated, because some of them might have huge operational costs. It is true that ICT generally used to automize the processes. While some of them are good in automation we need pay attention to other ones and wide the scope of understanding of ICT. We need involve this term not only in monitoring and automatization but also in other different fields like, HR, training and encouragement.

This fact is presumably making Kazakhstan remain in a lower position in the international ratings of fighting corruption. As a result of the paper, it is proposed to create and test an information and analytical system in the public authorities of Kazakhstan containing data in the following areas:

- Regulatory and legal support for combating corruption in the state and municipal service;
- Practical features of combating corruption;
- Memos and guidelines for combating corruption;
- Psychological tests (diagnostics of a tendency to corruption and other deviant deviations in behavior);
- Tests for knowledge of legal regulation of anti-corruption;
- Training.

6. References

[1] Anokhin S and Schulze W 2009 Entrepreneurship, innovation, and corruption *Journal of Business Venturing* **24**(5) pp 465-476 https://doi.org/10.1016/j.jbusvent.2008.06.001

[2] Zolotareva S and Mamonova I 2020 Anti-corruption in the civil service as a factor in increasing the efficiency of public administration *Social Integration and Development of Ethnocultures in the Eurasian Space* **1**(9) pp 132-137 https://articlekz.com/article/11828 (date accessed: 1.06.2020)

[3] Mukashev A 2020 How is the Fight Against Corruption in Kazakhstan Taken Place? *Central Asian Bureau for Analytical Reporting* https://cabar.asia/en/how-is-the-fight-against-corrup tion-in-kazakhstan-taken-place

[4] Transparency International 2019 *Corruption Perception Index*, https://www.transparency.org/en /cpi/2019/index/nzl

[5] Chene M 2012 Use of Mobile Phones to Detect and Deter Corruption Available from: https://www.gov.uk/dfid-research-outputs/

[6] Grönlund Å *et al* 2010 Increasing transparency and fighting corruption through ICT empowering people and communities *The Swedish Program for ICT in Developing Regions.* Available from https://spidercenter.org/files/2017/01/Spider-ICT4D-series-6-ICT-for-anticorruption.pdf

[7] Kossow N and Kukutschka R M 2017 Civil society and online connectivity: controlling corruption on the net? *Crime, Law, and Social Change* pp 1-18

[8] Sokolov S, Iliev T B and Stoyanov I S 2019 Analysis of Cybersecurity Threats in Cloud Applications Using Deep Learning Techniques *42nd International Convention on Information and Communication Technology, Electronics and Microelectronics (MIPRO)* pp 441-446 doi: 10.23919/MIPRO.2019.8756755

[9] Fauvel W 2021 Blockchain Advantages and Disadvantages Available from: https://medium.com/ nudjed/blockchain-advantage-and-disadvantagese76dfde3bbc0

[10] Krutz R L and Vines R D 2010 Cloud Security: A Comprehensive Guide to Secure Cloud Computing *Indianapolis: Wiley Publishing*

[11] Ansari Sh and Gaonkar R 2013 QR (Quick Response) Codes Application in Academic Libraries: Case study of KRRC (Central Library) *National Conference on "Changing trends in academic librarianship in electronic environment"*

[12] Stoyanov I S, Iliev T B, Mihaylov G Y, Evstatiev B I and Sokolov S A 2018 Analysis of the Cybersecurity Threats in Smart Grid University of Telecommunications and Post, Sofia, Bulgaria *IEEE 24th SIITME* pp 90-93 doi: 10.1109/SIITME.2018.8599261

[13] Otuoze A O, Mustafa M W and Larik R M 2018 Smart grids security challenges: Classification by sources of threats *Journal of Electrical Systems and Information Technology* **5**(3) pp 468-483 https://doi.org/10.1016/j.jesit.2018.01.001

Optimization of the operation of Dokan hydropower development using Cuckoo Search Algorithm

K Ahmad-Rashid[1]

[1] University of Sulaimani, Sulaymaniyah, Kurdistan, Iraq

E-mail: khalid.sd36@gmail.com

Abstract. In this paper one of the recently developed metaheuristic algorithms, the Cuckoo Search algorithm is used for the optimization of the operation of a large hydropower plant in Kurdistan, Iraq. The optimization problem is to realize an annual planned energy generation with monthly imposed fractions. The obtained results are excellent, nevertheless, there are some limitations of the algorithm determined by the initial level into the reservoir and a certain correlation between the type of the year, the starting level and the planned energy to be realized.

1. Introduction

The hydropower development (HPD) operation can be considered as an optimization problem that allows the achievement of the set objective by complying with a set of imposed restrictions [1]. The concept of "HPD management" or "operation" can refer to the following aspects:

- HPD size (one or more hydropower plants – HPPs);
- considered time horizon (online, daily schedule, medium or long term);
- hydrological conditions (known or probabilistic);
- concrete objective pursued (financial, technical, environmental, a combination of possibly contradictory requirements);
- internal (technological, maintenance) or external (legislative, market, etc.) restrictions.

According to the definition given by U.S. Bureau of Reclamation, the purpose of the hydropower management issue is related to the "selection of hydropower units and their loading so as to cover the demand for production and system services, using a minimum quantity of water for turbines" [2].

Regarding the optimization of the production of a single turbine, the aim is only to determine the used flow and not to cover a demand for services imposed for a given level in the reservoir. This leads to realize the maximum amount of energy from the turbined unit of water volume, to operate with optimal overall efficiency.

If the optimal operation of a HPP is analyzed, with all available units, the aim is to detect the configuration of units and their degree of load to cover the demand for services from HPP, with a minimum consumption of water from the reservoir.

An optimization problem involves defining a performance function, $f(\boldsymbol{x})$, depending on the vector $\boldsymbol{x} = x_1, x_2, \cdots, x_d$ of unknowns (decision variables). This function can be maximized or minimized, in the presence of equations that describe the functioning of the physical system considered and of various constraints, relationships in which the components of the vector $\boldsymbol{x}$ also appear. The solution of the optimization problem consists in finding the vector that satisfies the performance function, in accordance with the system equations and respecting all the imposed restrictions [3].

In the last period there have been many stochastic algorithms classified into heuristic and metaheuristic [4]. The heuristics algorithms are of the "tests and corrections" type, based mainly on the analyst's experience and the time available to find a satisfactory solution (not optimal guaranteed), by generating different solutions that explore the search space on its global scale, while metaheuristic algorithms (of higher level) have increased efficiency, due to the two characteristics: diversification and intensification, and by local search in the approach of better solutions. This leads to a multitude of convenient suboptimal solutions, with a much more limited search effort.

The most powerful metaheuristic algorithms are inspired by processes in the surrounding world and among them are evolutionary algorithms, which are algorithms that operate with populations of solutions. This, practically, began with the definition and use of genetic algorithms, between 1960 and 1970 (J. Holland, University of Michigan) and first papers related to it [5-7].

Other type of problems, than optimization of the operation related to hydropower, that can be addressed using optimization algorithms are:

- some studies can go, for example, from the determination of the impact of environmental flow on some small hydropower plants (SHPPs) production [8], to the impact of climate change and distributions of precipitation on hydropower production for a large country [9];
- use of dedicated software for assess the possibility to model the attachment of a HPP to existing dams [10], or presenting innovative concepts applied to recent hydropower projects [11];
- analysis of past energy production for determination of hydropower share in total energy generation [12], or of different statistical parameters [13,14], or prediction of share of hydropower in total energy generation [15];
- assessment of hydropower potential [16], or of pumped storage plants potential at a country level [17], or of hidden hydro related to empowering non-powered dams at a country level [18].

Related to optimization of the operation of HPDs, complex problems where the as the optimization of the electricity production of a HPP powered from a multiple use reservoir [19], or estimation of performance of a complex HPD with more reservoirs and associated HPPs and a pumped storage plant (PSP) [20] or the optimization of such complex HPDs [21], can be successfully addressed, where classical operational research methods can no longer face the complexity of the problem to be solved.

In this paper, the Cuckoo Search Algorithm (CSA) defined first time in 2009 [22] is applied for a large HPD from Iraq, Dokan, for the optimization of the operation of the related HPP.

2. Cuckoo search algorithm

To understand how the CSA was formulated, the following are some aspects that were Yang's source of inspiration in [22]. CSA is based on the parasitic reproductive behavior of some cuckoo species.

In fact, females of these species lay their eggs in the nests of other host species, having the ability to produce eggs similar in shape and color to those of the host, and even choosing the time of their laying (with shorter hatching time) - immediately after the parasitized bird.

If the host identifies eggs, either remove the foreign eggs from the nest, or leave it permanently and build another one. Otherwise, when the cuckoo's chick hatches first - it has the reflex to throw the host's eggs out of the nest, and if there are "stepbrothers" next to it - it adapts its chicks to theirs - to be favored for feeding by "adoptive parents".

This improved behavior of cuckoos in nature ensured the perpetuation of those species and suggested the schematization of CSA according to the following idealized rules:

- each cuckoo produces one egg at a time and lays it in a randomly selected nest;
- the nests with the most valuable eggs (solutions) are transferred to the next generations (iterations);

- the number of available nests is fixed, but a host can discover the foreign egg with a probability, $p_a \in [0; 1]$, in which case the host bird may either remove the egg from the nest or abandon it to build a completely new one in another location [22].

For simplification, the third rule can be approximated by a fraction, p_a, of those, n nests that can be replaced with new ones (containing new random solutions). The quality or adjustment of a solution in a maximization problem can be admitted as proportional to the objective function.

The basic equation of CSA by which a solution (cuckoo), i, is transferred from the current generation, t, to the next, t+1, is:

$$\boldsymbol{x}_i^{t+1} = \boldsymbol{x}_i^{t} + \alpha \cdot L, \tag{1}$$

where the second term in the right part corresponds to a random "flight" of the Lévy type, α is a parameter in the range [0…1]. This is a special type of distribution, with infinite average and variance, and the generation of implementation steps is done with Mantegna algorithm, [23], according to which:

$$s = \frac{u}{|v|^{1/\beta}}, \tag{2}$$

where u and v are normally distributed random numbers and β a parameter in the range [1…2].

The CSA provides that the fraction p_a of the total positions reached by equation (1) can be replaced by completely new solutions, randomly generated from the allowed domains for variables, avoiding convergence to local optimums. As iterations are in progress, the initial solutions (nests) improve their content and tend towards the overall optimum of the problem.

3. Case study

For materializing this study performed as related to the long-term optimization of a HPD, considering the probabilistic nature of the inflows, it was chosen one of the most important hydropower developments in Iraq, on Little Zab river, Dokan HPD, presented in figure 1.

Figure 1. Hydropower development Dokan.

Dokan development project has been initiated by the Council for Reconstruction and the General Directorate for Irrigation from the Ministry of Agriculture, at the end of years 1930. In year 1940, a British institute of consultancy has proposed to be realized a reservoir on Little Zab river; the institute also proposed several potential sites for dam location.

Although the project also foreseen the energetic use, the performance of a hydropower plant has been started only in 1975, the last of the 5 units being started for tests on 20th of November 1979. Equipment and installation have been carried out by the Technoprom company from USSR, but the construction part has been performed by a Yugoslavian company.

It results from the above mentioned that the completion of the whole project took 25 years and involved the participation of specialists from at least 5 countries of the world.

Under the design stage, the objectives for the project performance, were:

- protection against floods;
- use of water for irrigation;
- electric energy generation;
- pisciculture development;
- providing of favorable conditions for recreation.

The hydropower potential is turned to account through a hydropower plant, having an installed capacity of 400 MW in 5 identical units and generating, after commissioning, about 950 GWh/year.

Hydropower plant is built at the dam toe, being equipped with 5 Francis similar turbine with vertical axis, of 80 MW capacity, at net head of over 82 m and a flow rate between 95 and 111 m^3/s. The plant operates at a net head range between 60 and 95 m, with flow between 50 and 111 m^3/s. Maximum efficiency is achieved at heads over 77 m and flows over 90 m^3/s.

Within the initial project and upon the dam performance, in the dam body there were provided 4 penstocks for the HPP and one for irrigation. Subsequently it was assigned to the fifth unit, being constructively like the other ones. Each penstock has a diameter of 3.65 m and the maximum flow of 111 m^3/s. Access to the turbine is controlled by means of a butterfly valve weighing 76 t. The main housing is made of steel, with the spiral angle at over 350 degrees. The turbine runner has 16 blades, a diameter of 3.3 m and rated speed of 200 rot/min. Draft tube is also made of steel with a diameter increasing gradually from 3.3 m.

The turbined flow is discharged into a stilling basin provided at the downstream end with a sharp-crested spillway, over which the water is returned to the riverbed.

Dokan dam is a concrete arch dam, with a maximum height of 116.5 m and a crest length of 360 m. The base elevation is 399.5 masl, and the crest elevation is 516 masl. The upstream face is vertically built, and the thickness at main section varies according to third-degree equation, between 8.35 m at the crest and about 44 m at the base.

Though of rather low dimensions (height, crest length), Dokan dam achieves a huge reservoir, with a maximum level, under normal conditions, at elevation 511 masl and a minimum operating level at elevation 480 m masl, respectively. It is noticed that at elevation 511 masl the volume of water into the reservoir is of about 7 billion m^3.

At the elevation corresponding to the minimum operating level, 480 masl, the volume of water into the reservoir is still 1.45 billion m^3 and the surface exceeds 72.3 km^2. It results a live storage of about 5.4 billion m^3 achieved on a level difference of only 31 m.

The dead storage capacity is of 700 million m^3 and is situated under elevation 469 masl, where the water surface reaches 48 km^2.

Water supply for irrigation was the main objective upon the project drawing up and three tunnels were first carried out, with a total discharge of over 300 m^3/s. After the HPP construction, only a part of this arrangement remained as distinct work.

This supply system consists in a bank intake, a gallery that goes around the dam body and where two steel pipes of 2.28 m diameter and of about 160 m length each. The maximum flow likely to be conveyed is of 110 m^3/s. Each tunnel is provided with 3 safety and control valves, one at the upstream end and two at the downstream end. The water taken over from the reservoir is returned to the riverbed downstream of the stilling basin of the HPP, at an elevation of about 415 masl.

To apply CSA, the following were considered:

- capacity curve that was approximated with interpolation function "spline";

- specific energy production through $e(\overline{z_k}) = 0.002107(\overline{z_k} - 416)$, in [GWh/million m^3], where the coefficient of the right part is obtained considering the overall efficiency 77%, z represents free water level surface and 416 masl is the downstream level;
- the annual energy production from the project phase was admitted at $E_p = 950$ GWh/year, with the monthly productions imposed, E_k^*, $k = \overline{1, 12}$ and the average monthly multiannual inflows into the reservoir $Q_{in,k}$ (m^3/s), as in table 1;
- stored volumes at the end of each month k were limited: lower at V_{min} and upper at V_{max}, respectively, so that there is a reserve of water in the reservoir for the dry months and the possibility of flood control for the wet months (at the values in table 1);
- we also considered the driest year, and the wettest year, data presented in table 1: monthly inflows in Dokan reservoir, Q_d and Q_w, and monthly imposed energy generations in Dokan HPP, E_d and E_w, for driest year and wettest year, respectively.

Assuming that the initial volume, V_0, is known in the reservoir at the beginning of the year, the aim was to find the final monthly volumes in the reservoir $\boldsymbol{x}_k$, so that monthly energy production, E_k, be as close as possible to those required, E_k^*, if all the monthly effluent stock is processed.

The steps in CSA are: generation of new solutions keeping the current optimal solution, evaluation of new solutions, identification of weakest solutions and randomly replacing it, evaluation of new set of solutions and finding the optimal solution for each generation.

The performance function was chosen as:

$$\min \{f(x) = \sum_{k=1}^{12} (E_k - E_k^*)^2 \}, \quad (3)$$

where the energy generation is:

$$E_k = e(\overline{z_k}) \cdot D_k, \quad (4)$$

with $\overline{z_k} = (\boldsymbol{x}_{k-1}, \boldsymbol{x}_k)$ the average monthly reservoir level and $D_k(\boldsymbol{x}_{k-1}, \boldsymbol{x}_k)$ monthly stock for HPP operation:

$$D_k = \boldsymbol{x}_{k-1} - \boldsymbol{x}_k + V_{in,k}, \quad (5)$$

where $\boldsymbol{x}_0 = V_0$ for the first month, when $k = 1$.

Table 1. Data used in computation.

	1	2	3	4	5	6	7	8	9	10	11	12
V_{min} [mil.m^3]	1350	926	892	834	1159	1381	1932	2251	2302	2176	1978	1781
V_{max} [mil.m^3]	4300	3789	3303	4127	4940	5624	6737	7135	7043	6695	5816	4832
Q_{in} [m^3/s]	67	123	195	217	321	471	497	304	135	75	56	54
α_k [-]	0.1	0.1	0.09	0.09	0.08	0.07	0.07	0.06	0.07	0.08	0.09	0.1
E_k^* [GWh]	95	85.5	80.75	76	66.5	66.5	66.5	66.5	71.25	85.5	95	95
E_m [GWh]	83.05	61.64	62.01	55.06	47.18	47.17	37.46	53.71	76.56	118.89	150.18	116.25
Q_d [m^3/s]	16	23	39	108	125	146	162	88	31	16	12	13
E_d [GWh]	19.43	18.59	21.06	20.45	18.33	17.8	17.58	23.23	25.01	31.08	37.75	30.55
Q_w [m^3/s]	73	112	618	400	554	1510	814	515	256	162	84	91
E_w [GWh]	134.4	83.87	72.16	55.9	76.88	63.95	103.07	105.22	168.79	232.75	249.27	236.71

4. Results and discussions

To begin with, it was verified if CSA obtains the required energy of 950 GWh (the one from the project phase) for the mean hydrological year. The monthly values for the energy produced in HPP were calculated using monthly fractions, α, $E_k^* = \alpha_k \cdot E_p$, presented in table 1.

For verification, 10 runs were performed, and the results obtained can be seen in figure 2, a), with the variation of the volume in the reservoir at the end of each month. For these 10 runs were considered: $E_p = 950$ GWh, Q_{in} (from table 1) and the level at the beginning of the year was set up at 495 masl (correspond to the center of gravity of the reservoir as it split the live storage into two equal parts).

It can be observed that annual energy generation vary between 949.794 GWh/year (achieved after 896 generations) and 950.253 GWh/year (after 951 generations). For the graphical representation of realized energy generation, were chosen results from the 7th run, because of the smallest difference between the planned and realized energy generation. In this case, after 948 generations, energy generation was 950.023 GWh/year, the relative error being 0.0024%. Also, it can be observed that the planned energy and produced energy are very close, figure 2, b).

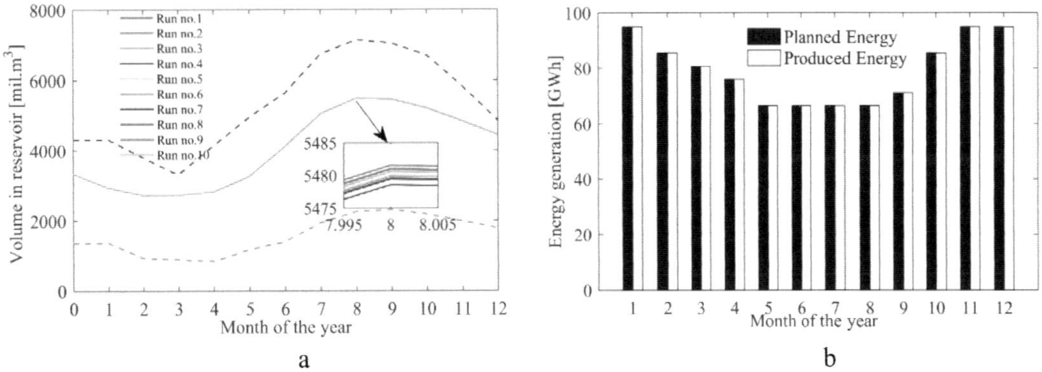

Figure 2. a) CSA results for 10 runs for the mean hydrological year, b) planned and produced energy.

There were computations for monthly energy generation for mean hydrological year for three different water levels, respectively known volumes at the beginning of the study period, namely: Z_{01} = 480 masl, the minimum operating level (mOL), an intermediate level, Z_{02} = 495 masl and Z_{03} = 500 masl, the maximum operating level (MOL). The obtained results are presented in figure 3.

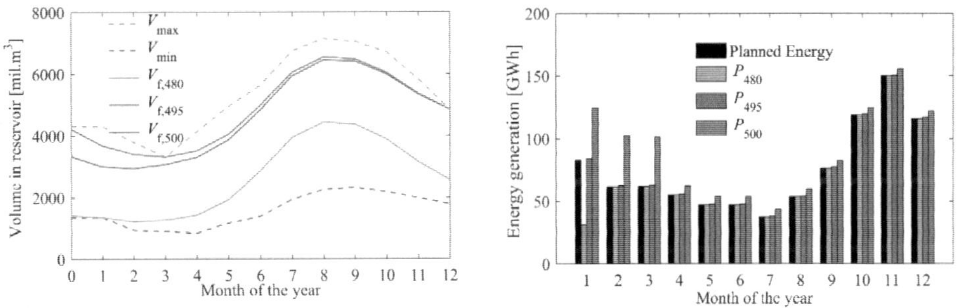

Figure 3. Results for the mean hydrological year, for three imposed levels.

It can be easily observed that for the intermediate level planned energy is achieved month by month. Starting with the mOL, only in first month it was not possible to achieve the planned energy as it was no available water in the reservoir. Starting with the MOL, due to upper limitations of volumes into the reservoir, produced energy is larger than the planned one as it was compulsory to turbine more water than for realizing the imposed energy.

The results of the computations for the driest and for the wettest years are presented in figure 4 and figure 5.

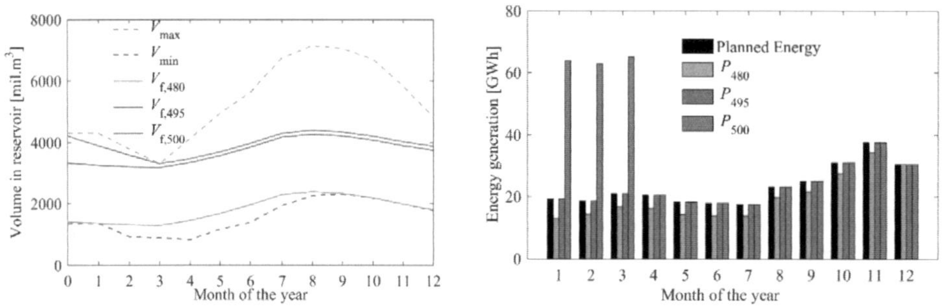

Figure 4. Results for the driest year, for three imposed levels.

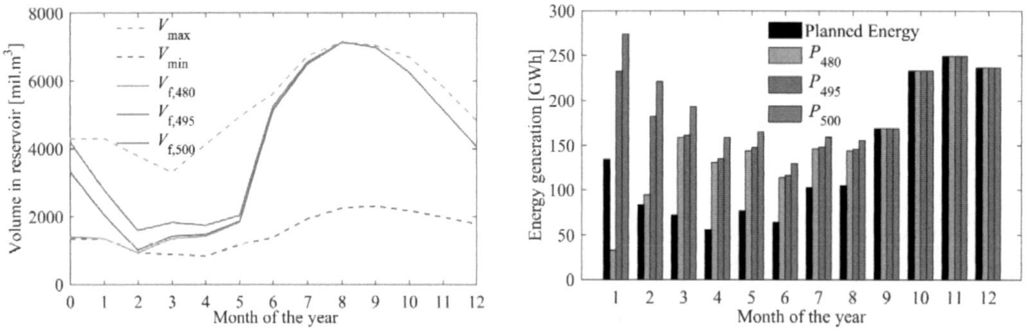

Figure 5. Results for the wettest year, for three imposed levels.

For the driest year, planned energy can be realized starting with the intermediate level; very close starting with mOL and exceeded for the first 3 months and realized for the rest starting with MOL.

For the wettest year, planned energy can be realized in all alternatives of starting level for the last four months. Excepting the alternative of starting with mOL where energy generation in first month was well below and in the second month close to the planned energy, for first eight months, energy production was greater than the planned energy and relatively close one from the other.

5. Conclusions

Many papers demonstrates that metaheuristic algorithms can be successfully used for the optimization of the operation of HPDs. In this paper, CSA was used at an impressive HPD, for realizing a planned annual energy generation, month by month.

There were 10 computations for an intermediate level in the reservoir for the mean hydrological year and the obtained results were excellent. Then, for three levels at the beginning of the year, there were computations for the mean hydrological year, for the driest and for the wettest year. Concluding the discussion from the previous section: all volumes variations respected the imposed min and max limits and the planned monthly energy was realized excepting the situations when starting from the MOL it was compulsory to turbine more than for realizing the planned energy (first three months for the driest year) and the situations when the inflows correspond to the wettest year when for energy production was larger than the planned one, excepting last four months where it was equal for all starting level alternatives.

The overall conclusion of the paper is that, for the optimization of the operation of a HPD defined in terms of realizing a planned energy generation with monthly imposed fraction, CSA gives good results.

6. References

[1] Tică E I 2019 Contributions to the optimization of hydropower developments operation with evolutionary algorithms - PhD thesis (Bucharest)

[2] Harpman D A 2012-2013 Advanced Algorithms for Hydropower Optimization - Technical Report (Denver: Colorado)
[3] Popa R and Popa B 2003 Optimization of the operation of hydropower developments (Bucharest: Technical Publishing House) p 115
[4] Yang X S 2008 Nature – Inspired Metaheuristic Algorithms, Lumiver Press
[5] Holland J 1975 Adaptation in Natural and Artificial Systems, Univ of Michigan Press – Ann Arbor
[6] Goldberg D E 1989 Genetic Algorithms in Search, Optimization and Machine Learning, Addison – Wesley, Reading, Mass
[7] Michalewiez Z 1992 Genetic Algorithms + Data structures = Evolution Programs, Springer – Verlag, N, Y.
[8] Popa F, Dumitran G E, Vuta L I, Tica E-I, Popa B, Neagoe A 2020 Impact of the ecological flow of some small hydropower plants on their energy production in Romania. *Journal of Physics: Conference Series*, **1426**, 012043
[9] Bulut U, Sakalli A 2020 Impacts of climate change and distribution of precipitation on hydroelectric power generation in Turkey. *IOP Conf. Ser.: Mater. Sci. Eng*, **1032**, 012043
[10] Moldoveanu A, Tică E-I, Popa B, Popa F 2017 Assessment of the possibility to model a Toe Dam hydropower plant using dedicated software. *International Conference on ENERGY and ENVIRONMENT (CIEM)*, pp 260-264 doi: 10.1109/CIEM.2017.8120781
[11] Dadu V, Dadu A, Frunza D, Catarig G, Popa F, Popa, B 2017 Innovative Concepts Applied to Recent Small Hydropower Plants, *Energy Procedia*, **112**, pp 426-433
[12] Neagoe A, Tică E-I, Diminescu M, Popa B, Cismas S C 2019 Analysis of hydropower ratio from total energy production in Romania, *E3S Web Conf.*, **85** 06009 https://doi.org/10.1051/e3sconf/20198506009
[13] Neagoe A, Tică E-I, Popa F, Popa B 2020 Change point detection in recent hydropower generation in Romania, *IOP Conf. Ser.: Mater. Sci. Eng.* **1032** 012045
[14] Neagoe A, Tică E-I, Achim A V and Popa B 2021 Trend Detection in Seasonal Inflows and the Influence on Energy Generation in Goleşti Hydropower Plant. *The Journal of CIEES*, **1**(1) pp 40-49 https://doi.org/10.48149/jciees.2021.1.1.1
[15] Neagoe A, Tică E-I, Popa B, Popa F 2020 Prediction of Hydropower Ratio from Total Energy Generation in Romania *TEM Journal* **9**(1), pp 402-406
[16] Moldoveanu A, Galie A, Moldoveanu M, Popa F, Tică E-I, Popa B 2016 Assessment of small hydropower potential for the Olanesti River using advanced software techniques. *IOP Conf. Ser.: Mater. Sci. Eng.* **163** 012022
[17] Popa F, Popa B, Popescu C 2017 Assessment of Pumped Storage Plants in Romania. *Energy Procedia*, **112** pp 473-480 https://doi.org/10.1016/j.egypro.2017.03.1122
[18] Vuta L I, Dumitran G E, Popa B, Diminescu M A, Tică E-I 2019 Hidden hydro related with non-powered dams in Romania. *Proceedings of CIEM* pp 413-417 doi: 10.1109/CIEM46456.2019.8937680
[19] Popa B, Tică E-I, Popa F 2020 Optimization of Electricity Production of a Hydropower Plant Powered from a Multiple Use Reservoir, Using an Adapted SSO Algorithm. *U.P.B. Sci. Bull., Series C,* **82**(3), pp 291-299
[20] Tică E-I, Popa B, Popa R 2017 Annual Performance Estimation of a Multireservoir System Including a Pumped Storage Plant, for the Mean Hydrological Year. *Journal of Energy Engineering*, **143**(6) DOI: 10.1061/(ASCE)EY.1943-7897.0000489

[21] Popa R, Popa F, Popa B, Zachia-Zlatea D 2010 Optimization of the weekly operation of a multipurpose hydroelectric development, including a pumped storage plant. *IOP Conf. Ser.: Earth Environ. Sci.* **12** 012118

[22] Yang X S, Deb S 2009 Cuckoo Search via Lévy flights, *World Congress on Nature & Biologically Inspired Computing (NaBIC)*, pp 210-214, doi: 10.1109/NABIC.2009.5393690

[23] Mantegna R N 1994 Fast, accurate algorithm for numerical simulation of Lévy stable stochastic processes, *Physical Review*, **49**(5), pp 4677-4683 https://doi.org/10.1103/PhysRevE.49.4677

Optimization of a complex hydropower development operation using HEC-ResSim

E I Tică[1], K Ahmad-Rashid[2], O V Sima[1], F Popa[3], O Nedelcu[3] and B Popa[1]

[1] Department of Hydraulics, Hydraulic Machines and Environmental Engineering, Faculty of Energy Engineering, University Politehnica of Bucharest, Romania
[2] Electrical Engineering Department, College of Engineering, University of Sulaimani, Sulaymaniyah, Kurdistan, Iraq
[3] Department of Electronics, Telecommunication and Energy Engineering, University Valahia of Târgoviște, Târgoviște, Romania

E-mail: bogdan.popa@upb.ro

Abstract. In this paper HEC-ResSim is applied for a complex hydropower development formed by five reservoirs and related hydropower plants. There were considered characteristics of five existing hydropower developments in Romania, for which three reservoirs are with annual regulation and two with daily regulation. The objective function was the realization of a planned energy generation for one year (the mean hydrological year). Obtained results are very close to those obtained applying linear programming, a revised simplex algorithm.

1. Introduction

U.S. Army Corps of Engineers US Army Corps of Engineers offer a performant tool, Reservoir System Simulation (HEC-ResSim) software, based on which decisions can be made in real time operation. HEC-ResSim is a decision support tool that meets the needs of modelers performing reservoir project studies as well as meeting the needs of reservoir regulators during real-time events. This can be used by all hydropower developments users applied for one or more reservoirs [1].

There are papers with HEC-ResSim applied for different arrangements for hydropower developments to optimize the hydropower production, also to operate the hydropower development (HPD) considering the climate change scenarios [2-4]. For example, in [5] it was used to simulate the reservoir operation for the optimization of multiple reservoir system placed on Nam Ngum River basin. The main goal was to maximize the hydroelectric power generation. Operational behavior of the reservoir Dokan, Iraq, was analyzed in [6], where the simulation model was based on monthly flows. The results, in terms of storage, area, elevation, were compared with observed data for the actual operation. Thus, it was observed a good correlation between the outflows from the reservoir calculated from the model with and observed and between the area and volume curves of the reservoir with the water level in the reservoir.

In paper [7] was presented a simulation model in HEC-ResSim for water management in which were considered water loss by evaporation in El Cuchillo reservoir. For the simulation, two scenarios were studied to save more water from evaporation.

In other paper [8] was presented a simulation model for the evaluation of reservoir operation flexibility for Yuvacik reservoir under variable hydrological conditions.

Also, in a previous paper [9] was tested the applicability of the HEC-ResSim in a simple optimization problem formulated to achieve an established generation plan as fraction from annual average energy.

In this paper a simulation-based approach is presented for the optimization of a complex HPD system formed by five reservoirs and related hydropower plants (HPPs) implemented in HEC-ResSim. The optimization problem was formulated to achieve the planned electricity generation, month by month, to ensure the volumes of protection against floods, but also the flows required by consumers. The restrictions were formulated with respect to the constructive characteristics of the works, but also to minimize the losses of productive energy in the system. The simulation results from HEC-ResSim were compared with those obtained in [10] applying linear programming, a revised simplex algorithm, for the same complex system of reservoirs and related HPPs.

For the optimal formulation and operation of all HPPs in a complex development or in a river basin, the size of the hydropower development or the river basin, the number and volumes of reservoirs, the configuration of the river network participating in the production are largely considered of energy, of the type of arrangements (large reservoirs and HPPs, HPPs cascade with small reservoirs or on derivation, mixed structure with peak HPP and, respectively, in waterfalls, etc.).

For such problems, it is complicated or even impossible for the solution found to simultaneously ensure both the coverage of the required service demand and the condition of its realization with a minimum water consumption in the system. In the following works [11-15], topics related to the energy production for multipurpose reservoir and multiple water uses were approached.

2. Complex hydropower development

To illustrate the main notions involved in the deterministic optimization model and performing simulations in HEC-ResSim, the HPD shown in figure 1 was used, composed by five reservoirs and related HPPs.

The development includes three large reservoirs (1, 2 and 5) with possibilities for annual regulation and associated peak HPPs: 1, 2 and 5. The reservoirs 3 and 4 are allowed with small volumes, buffer reservoirs, with daily regulation. HPP 3 can be operated at peak as well as at base. HPP 4 and HPP 5 are located on the penstocks 1 and 2, respectively. On the penstocks 1 and 2 are located HPP 4 and HPP 5. The inflow stocks in the large reservoirs are referred as a_1, a_2 and a_5.

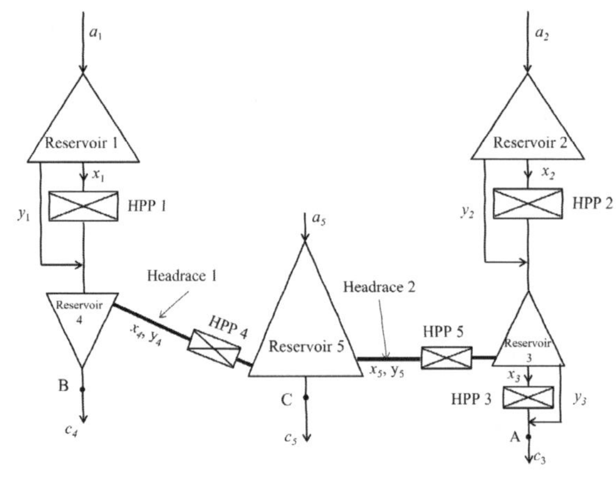

Figure 1. HPD scheme for exemplification [10].

In downstream sections A, B and C, stocks c_3, c_4, c_5 are required for other users. In the deterministic context, it is assumed that the following are known: reservoirs capacity curves, tributary and downstream demand stocks, inflows and capacities installed in HPPs, specific energy generation and capacities available in the HPPs (as head functions or volumes in reservoirs), the maximum transport capacities of the supply galleries, the monthly protection volumes in the large reservoirs, necessary for flood control, the minimum monthly volumes allowed in large reservoirs, the maximum monthly number of hours on HPPs (how long they can operate at peak load), the annual energy generation planned of peak HPPs and its monthly fractions, etc.

Most optimization problems involve the definition of the following elements: the stage variable, state variable and decision variable; performance function(s) and system operating restrictions.

In this paper, it is admitted as stage variable - the month, and the analysis horizon - the year, so the long-term policy is targeted. A selection alternative for the state variables could be the volumes in large reservoirs, forming the three-dimensional vector **V**. On a monthly time-step, buffer reservoirs 3 and 4 are considered without storage possibilities.

As decision variables are admitted the stocks used at the peak load and, respectively, those turbined out of the peak or spilled, forming vectors with 5 components, **X** and **Y**.

Usually, a system objective function is written as [10]:

$$\max_{\mathbf{X},\mathbf{Y}}(\text{or }\min)\{F = \textstyle\sum_{n=1}^{N} \alpha_n \cdot f_n(\mathbf{V}_n, \mathbf{A}_n, \mathbf{D}_n, \mathbf{X}_n, \mathbf{Y}_n) + \alpha_{N+1} \cdot \phi_{N+1}(\mathbf{V}_{N+1})\} \tag{1}$$

where: n is the stage index (monthly step); N is the number of stages on the annual horizon; **A** and **D** are vectors of inflowing stocks in large reservoirs and on river basin difference; f is the objective function followed in the stage n (depending on the state, decision and inflow stock vectors); ϕ introduces possible benefits (costs) beyond the horizon N (depending on the final state); and α are possible value equivalence coefficients (if f is expressed in terms of financial units).

For the model formulation were considered reservoirs 1, 2 and 5 have considerable volumes which allows the annual regulation of stocks, and reservoirs 3 and 4 operate to carry out the daily regularization of flows from upstream HPPs.

The analyzed HPD is a development with multiple uses: hydropower, ensuring the flows required for the water supply of consumers, protection against floods, but also the transfer of water between adjacent river basins.

The following input data was allowed to configure this system: average monthly inflows, minimum inflows required by consumers, monthly energy demand, inflows and powers installed in HPPs, minimum monthly volumes allowed and monthly volumes of flood protection, maximum monthly number of operation hours per HPP, how long they can operate at peak, but also data on installed capacity and inflow, efficiency and head at the normal retention level from each reservoir.

The objective of the optimization problem is to achieve month by month, energy generation, to ensure the volumes of protection against floods, but also to ensure the flows required by consumers, respecting the imposed restrictions.

To achieve the energy function, the annual generation of planned energy and the monthly fractions of its realization (α_k) were considered, values that are established by the system dispatcher in accordance with the position of the month in the year.

For the analyzed complex HPD were considered the constructive and hydrological data of some of the most important HPDs from Romanian National Power Grid, even if there are not really interconnected (they are not hydraulically coupled):

- Reservoir/HPP 1 – reservoir Vidra, HPP Lotru-Ciunget;
- Reservoir/HPP 2 – reservoir Vidraru, HPP Vidraru;
- Reservoir/HPP 5 – reservoir Izvorul Muntelui, HPP Stejaru.

For the mathematical formulation of the problem in deterministic terms were:
- a stage variable that represents the monthly interval k, k=1, 2.., 12, during a year;
- a state variable represented by the volume in the reservoir at the beginning of a month, V_k^i;
- a decision variable that represents the volume in the reservoir at the end of a month, V_k^f.

To fulfill the system state transformation relationship, the volume at the end of month k, will become the initial volume for month $k + 1$, fulfilling the balance equation:

$$V_k^f = V_k^i + a_k - x_k - c_k - y_k, \tag{2}$$

where V_k^f is the volume in reservoir at the end of the month, V_k^i is the volume in reservoir at the beginning of the month, a_k – is the unregulated tributary stock, x_k – turbined volume in the associated HPP at the peak load, c_k – the volume extracted for other consumers, y_k – turbined volume in the associated HPP out of the peak load or spilled.

If reservoir volume at the beginning of the month is known, as well as the stocks of tributary and extracted for other uses, the condition of the reservoir at the end of the month is determined by the volumes used at peak, x, and turbine or discharged outside peak hours, y.

The following restrictions about volumes for reservoirs 1, 2 and 5, appears in the operation of the system:

$$\begin{gathered} V_1^{\max} - V_1 \geq V_1^i + a_1 - x_1 - y_1 \geq V_1^{\min} \\ V_2^{\max} - V_2 \geq V_2^i + a_2 - x_2 - y_2 \geq V_2^{\min} \\ V_5^{\max} - V_5 \geq V_5^i + a_5 + x_4 - x_5 + y_4 - y_5 - c_5 \geq V_5^{\min} \end{gathered} \tag{3}$$

where min indicate the minimum permissible volume, max – the maximum possible volume and V – the volume required for flood control. The minimum permissible volume and V may be different from month to month.

For each HPP the following data related to the installed capacity, P_i, installed flow, Q_i, and gross head, H_{gr}, were considered:

- HPP 1 – P_i = 510 MW, Q_i = 80 m3/s, H_{gr} = 809 m;
- HPP 2 – P_i = 220 MW, Q_i = 90 m3/s, H_{gr} = 324 m;
- HPP 3 – P_i = 40 MW, Q_i = 150 m3/s, H_{gr} = 38,8 m;
- HPP 4 – P_i = 20 MW, Q_i = 80 m3/s, H_{gr} = 36,4 m;
- HPP 5 – P_i = 210 MW, Q_i = 178 m3/s, H_{gr} = 149 m.

3. Methodology

The operation simulation model of the complex HPD implemented in HEC-ResSim consists of achieving month by month the planned energy generation, to ensure the volumes of protection against floods and flows required by consumers in HPDs with long-term regulation reservoirs 1, 2 and 5.

For reservoir 1, compared to the mean inflow of 18.7 m^3/s, at which the average annual generation is 1078 GWh, the mean inflow of 8.24 m^3/s was considered. In the case of reservoir 2 with the mean inflow 19.7 m^3/s and the average energy 400 GWh, the mean inflow of 18.4 m^3/s was allowed, and in the case of reservoir 5 with the mean inflow 42 m^3/s and the energy generation 434.5 GWh, the mean inflow of 53.5 m^3/s was considered.

Because in the reservoir whose power plant provides approx. 56% of the energy in the average year, very small tributary flows were imposed compared to those in the average year (8.24 m^3/s compared to 18.7 m^3/s), the annual planned peak energy on the system was set to be 1200 GWh , and the volumes at the beginning of the year, V_i, i = 1, 2, 5, were chosen: V_1 = 260 mil.m^3 – for reservoir 1, V_2 = 315 mil.m^3 – for reservoir 2 and V_5 = 800 mil.m^3– for reservoir 5.

For simulation, HEC-ResSim requires the creation of a scheme configuration that is represented in figure 2 and the choice of the time step for the calculations.

The configuration of the complex system of HPD was done based on the physical and operational elements of the reservoirs and the alternatives for operation simulation will be created. Using the Reach Tool it is possible to impose on the program the river sections for which the hydrological data are known, and for the entry / exit points of the system they will be created with the help of the Junction Tool. The water supply of the domestic / industrial consumers was ensured by creating three supply penstocks in reservoir 3, 4 and 5, these were created through the Diverted Outlet tool. After introducing all the physical components necessary for a simulation within the HEC-ResSim program, it is necessary to create a set of operations that will characterize the operation of the development.

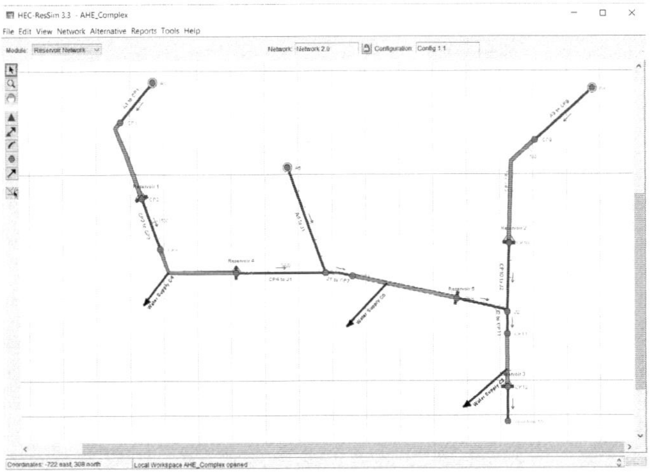

Figure 2. Representation of the system in the Reservoir Network module.

A set of operations is represented by three basic characteristics: zones, rules and identification of the guidance curve. The areas are operating subdivisions of the reservoirs. Each area is characterized by a curve that indicates its upper limit. After creating an operation, HEC-ResSim establishes a predefined set of zones, these being: Flood Control, Conservation and Inactive. The idle zone is a special area in the set of operations, it represents the dead volume and the, so called, iron reserve.

To achieve the planned energy generation, knowing the annual energy generation and the monthly fractions of its realization presented in table 1, a rule will be introduced for the entire HPD system.

Table 1. Monthly fractions and planned monthly energy generation.

	1	2	3	4	5	6	7	8	9	10	11	12
α_k	0.1	0.1	0.09	0.09	0.08	0.07	0.07	0.06	0.07	0.08	0.09	0.1
E_k^p [GWh]	120	120	108	108	96	84	84	72	84	96	108	120

To add the rule, select the area that is represented by the guidance curve, in this case Flood Control and the Add New Rule command will specify the type of rule Hydropower - System Schedule and the component that is usually restricted, in this case HPP 5. The rule editor will specify the monthly energy production and the plants participating in the energy generation.

To ensure the minimum flow rates required by consumers, it is necessary to introduce a rule for the Controlled Outlet C5 component of the Release Function type. It will specify the minimum flows required by consumers depending on the time, limit and type of interpolation of values.

HEC-ResSim program describes an alternative as being designed from a series of data on the studied arrangement, control specifications, a set of operations, a description of the initial conditions (arrangement state before the start of the simulation) and recording of all inflow data in reservoir.

Simulation model of the complex system of HPD have the purpose to isolate the output analysis from the model development process. Once the reservoir model is complete and the alternatives defined, the simulation module is used to configure the simulation.

In this module, a time interval and a step for an alternative or group of alternatives are specified. After a successful simulation, the results can be analyzed, with the possibility to perform additional revisions for a better evaluation of operations.

In this case, the period in which the simulation takes place was 01.01.2021 to 01.01.2022, using the alternative created in the previous Alt 2.0 module. After creating the simulation, the Compute button will be activated, after completion it can see the results both tabularly and graphically, it can be created reports and a detailed analysis, if State Variables have been created, using the Debug console.

4. Results and discussions

The results obtained were compared with those presented in [10], where a deterministic model for optimizing the long-term operation of a complex system of reservoirs and HPPs was resolved using the Simplex algorithm.

In table 2 are centralized the values for the final monthly volume from the large reservoirs 1, 2 and 5 obtained using HEC-ResSim program, respectively Simplex algorithm.

Table 2. Final monthly volumes, V_f, for reservoirs 1, 2 and 5, in [mil.m^3].

			1	2	3	4	5	6	7	8	9	10	11	12
Res. 1	Simplex	V_f	246.8	232.0	227.3	250.8	294.4	319.0	324.0	331.8	349.0	326.0	327.2	317.5
	ResSim	V_f	233.6	217.7	224.4	247.7	283.8	315.6	324.0	336.1	334.9	333.2	322.6	304.9
Res. 2	Simplex	V_f	291.2	258.8	223.4	247.9	322.2	405.6	495.0	484.1	480.0	474.7	460.2	439.5
	ResSim	V_f	252.7	209.7	223.4	298.8	408.0	505.4	537.0	527.3	524.1	518.2	494.8	448.9
Res. 5	Simplex	V_f	714.0	607.1	589.1	722.0	795.9	1110.0	1150.0	1191.2	1137.2	1170.5	1077.0	971.7
	ResSim	V_f	823.6	767.2	777.5	854.7	074.2	1097.7	1149.3	1164.8	1179.9	1152.2	1115.7	1082.9

Using *Reservoir Summary Report,* it can be centralized data about operation of HPPs throughout the studied period: head (H), energy generation (E) and turbined flow (Q), which are presented in table 3.

Table 3. Results obtained for HPPs 1, 2 and 5.

		Mean	Max	Min
HPP 1	H [m]	796.91	808.15	779.23
	E [MWh]	1019.54	11766.12	0
	Q [m^3/s]	6.98	80.00	0
HPP 2	H [m]	303.01	331.39	247.3
	E [MWh]	782.93	5244.25	0
	Q [m^3/s]	14.12	89.37	0
HPP 5	H [m]	144.57	155.66	127.66
	E [MWh]	1216.17	4939.86	0
	Q [m^3/s]	46.61	178.00	0

Figure 3 shows the variation of the volumes at the end of the month stored in the reservoirs 1, 2 and 5 during a year, determined with the HEC-ResSim simulation program and the variation obtained by the Simplex algorithm [10], together with the lower and upper limits, imposed on the final monthly volumes stored in the reservoir for the initial volumes specified above. As can be seen, the variations obtained in the HEC-ResSim simulation program differ monthly from the optimal solution obtained by the Simplex algorithm, but they fall within the allowed variation band.

The final volumes for all reservoirs are higher than the initial ones, it is found that the simulation program favored the balancing of stocks in reservoirs (the imposed rules allow the simulation program to balance stocks in reservoirs to configure the system for a new model simulation with different operating rules), compared to the Simplex procedure which stimulated water conservation in reservoirs 1 and 5.

The Release Schedule function imposed an operating rule for reservoirs 3, 4 and 5, to ensure the necessary flows for domestic / industrial consumers, they were represented in figure 4. It can be seen the fulfilment of the flow insurance condition.

The monthly planned monthly energy generation and that realized using HEC-ResSim, are shown in table 4.

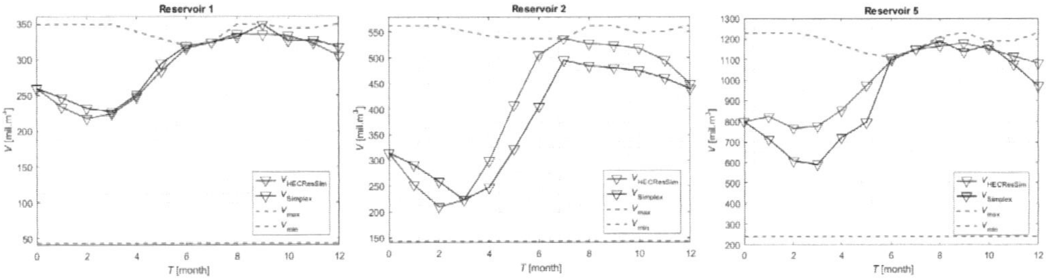

Figure 3. Monthly volumes variation in reservoirs 1, 2 and 5.

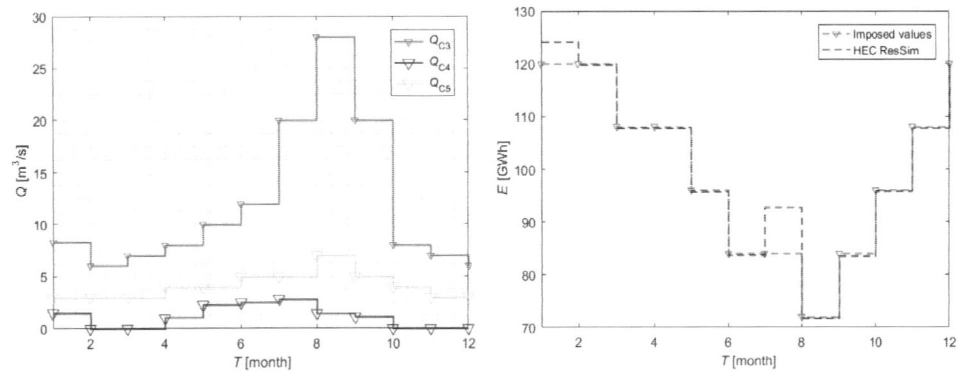

Figure 4. Required inflows and time variation of monthly energy generation.

Table 4. Planned monthly energy generation and values realized using HEC-ResSim, in [GWh].

	1	2	3	4	5	6	7	8	9	10	11	12
Imposed values	120	120	108	108	96	84	84	72	84	96	108	120
HEC-ResSim	124.19	119.84	107.78	107.78	95.75	83.77	92.71	71.78	83.51	95.86	107.87	119.88

In table 4 are presented related data about energy generation for each month of the analyzed year. On the second line of the table are shown numerical values of the imposed (planned) energy, month by month, for the analyzed HPD. On the last line of the table, the values obtained in HEC ResSim were presented. It can be easily observed that the two sets of values are very close.

From these data it is observed that only in January and August the specified energy generation is exceeded, for January the cause is the simulation model that requires the introduction of a turbine flow before the start of the simulation, and the surplus in July is due to the restriction of maximum volumes allowed in reservoirs.

Annual energy generation calculated using the Simplex algorithm [10] was 1233.6 GWh with relative error, compared to annual imposed energy 1200 GWh, of 2.8 %. Annual energy generation calculated using the HEC-ResSim simulation model was 1210.7 GWh, so the relative error related to linear programming, a revised simplex algorithm, is only 0.89 %.

If the difference between the imposed value and the calculated ones is analyzed from a hydropower point of view, it results that the HEC-ResSim simulation program led to an insignificant difference.

5. Conclusions

Through the HEC-ResSim program, a simulation model was developed for an optimization problem related to a complex HPD.

The obtained results were compared with those in the bibliographic reference [10], from where the problem, data and hypotheses were taken. The results are convincing enough to motivate interest in using the program to solve problems related to optimization of the operation of hydropower facilities.

A difference in time variation was observed for the final monthly volumes in the reservoirs, calculated using HEC-ResSim software and the variation calculated with the reference solution which stimulated water conservation especially in reservoirs 1 and 5.

This difference can be generated by the time step for which the simulation was performed, because the program allows as a simulation step a maximum of 1 day. Thus, the program accurately calculates the flows entering the reservoir compared to the reference solution that was made with a monthly time step. This is an advantage because, with this calculation, the result is closer to the actual operation.

Another reason that could generate this difference is that the program tends to balance the stock of reservoirs, to reach the end of the simulation period at the initial volume in accordance with the principles of operating a reservoir with annual adjustment.

The simulation model complies with all the operating conditions imposed on the HPD system such as: monthly energy generation, insuring flows to other consumers, protection against floods.

Compared to the reference solution, the value of the annual energy generation achieved by this method was the smallest relative error compared to the required value. Therefore, it can be concluded that the program optimizes the layout to produce the required energy demand using a minimum volume considering the rest of the operating restrictions.

The results of the simulation model developed in HEC-ResSim are very close to those obtained from previous studies for the same arrangement.

In conclusion, this paper demonstrates the simplicity and efficiency of using the HEC-ResSim simulation program to solve various problems of optimization of the operation of HPDs, compared to the development of dedicated software, the results being enough accurate.

6. References

[1] US Army Corps of Engineers, https://www.hec.usace.army.mil/software/hec-ressim/
[2] Lin N M and Rutten M 2016 Proc. Eng. 154 pp 1376-1384
[3] Shrestha A et al 2021 J Clean. Prod. 279 123483
[4] Hughes D et al 2021 Env. Mod.& Soft. 138 104980
[5] Sorachampa P et al 2020 Energy Rep.: Conf. Series 6 KS6ZC
[6] Sulaiman S O et al 2021 I. J. Des. Nat. Ecodyn. 16 160308
[7] Stella J M and Gordon P 1 2019 Atmos. 32
[8] Uysal G et al 2013 C. Hydrol.Ch. Res. Plan. Manag.
[9] Tică E I et al 2020 J Eng Sc
[10] Popa R and Popa B 2003 Optimization of the operation of hydropower developments (Bucharest: Technical Publishing House) p 115
[11] Neagoe A et al 2021 J CIEES 1 pp 40-49
[12] Wang C and Chen S 2019 Hindawi Comp. 1534598
[13] Ribeiro A F et al 2012 Elect. P Syst. Res. pp 121-129
[14] Kling H et al 2014 J of Hydrology pp 17-43
[15] Thushara de Silva M and Hornberger G M 2019 J of Hydrology 25 100624

AUTHOR INDEX